Advances in Aquatic Biology and Toxicology

Advances in Aquatic Biology and Toxicology

Edited by

Dr. Apoorva P. Hiray
Dr. B.S. Yadav

ADHYAYAN

Adhyayan Publishers & Distributors
New Delhi-110002

Published by
ADHYAYAN PUBLISHERS & DISTRIBUTORS
4378/4 B, 105, J.M.D. House, Murari Lal Street
Ansari Road, Darya Ganj, New Delhi-110002
Ph. : 011-23263018, 011-23277156, Fax : 011-23280028
E-mail : adhyayanpublishers@yahoo.com

Advances in Aquatic Biology and Toxicology

Edition : 2010
ISBN : 978-81-8435-221-4

Printed in India

Published by Harish Chandra Yadav for Adhyayan Publishers & Distributors, Laser Typesetting at Net Graphics and Printed at Tarun Offset Printers, Delhi.

Preface

Water is essential for the ecosystem and has been termed as the elixir of life. The role of water in nature is unique not only from the point of human consideration, but numerous of those organism which makes aquatic medium their abode. Man uses water for many purposes such as drinking, irrigation fisheries, industrial proceses, transportation and waste disposal. Industries and forms have already reduced the availability of water to below its minimum per capita daily requirements.

Nowadays water bodies have become dustbin causing severe water pollution due to rapid growth of population abundant use of fertilizer in agriculture as well as untreated discharge of domestic sewage. This has resulted in the degradation of physico-chemical as well as biological quality of water. Such deteriorating environment of the fresh water ecosystem is a matter of great concern. The polluted state of natural water resources has led to a steady decline in aquatic productivity and it has also adversely affected the aquaculture strategies.

The problem of water pollution has drawn global attention in recent years to findout the ways to sustain and manage environment. It has challenged the environmentalists, technologists, politicians and also socio- economic reformers.

The aim of this book is to provide the readers materials on the subject in a lucid and readable from so as to enable the environmentalists, biologists and even the common man to understand the subject properly.

The book have been divided into two parts. The 1st part entitled 'Physico-chemical Studies' containing articles and presents the latest researches on various dimensions of water parameters and 2nd part entitled 'Toxicological Studies' containing latest research papers on various aspects of environmental pollution.

We are sure that this book would be found quite useful to the researchers, organizations and agencies involved in the water management sector.

Words cannot adequately express our indebtedness to honourable Samaj Shri Prashant Dada Hiray, General Secretary Mahatma Gandhi Vidyamandir Malegaon for his sustained guidance, help and encouragement have been a tower of strength to our work. We are also thankful to Shri Adavay, P. Hiray Director NDCC bank Malegaon for encouragement and support.

Thanks are greatfully rendered to the Principal, MSG college for his valuable suggestions. We are also thankful to our friends who always encouraged us to complete this work.

Our thanks are also due to Shri Harish Chandra Yadav, Adhyayan Publishers and Distributors for publishing this book.

Dr. Apoorva P. Hiray
Dr. B.S. Yadav

Contents

PART I

Physico-Chemical Studies

1

Monthly variations in DO, BOD, and COD parameters in Godavari river at Nashik

Resham Bhalla
B.S. Yadav

ABSTRACT

The study was conducted to determine the monthly mean variations in DO, BOD and COD parameters at three locations (Station A, Station B and Station C) during November 2001 to October 2002 and November 2002 to October 2003. The depletion of DO values, due to discharge and degradation of wastes at three stations indicated that the Godavari river was polluted. Low DO content in water resulted into high BOD values, the basic aim is to degrade the organic matter. The BOD and COD values were observed more during monsoon than winter and summer.

Keywords : *Godavari river, DO (Dissolved Oxygen), BOD (Biochemical Oxygen Demand), COD (Chemical Oxygen Demand), Mean Values.*

Introduction

Water quality is measured in terms of parameters like DO, BOD and COD. The Gadavari is the largest river of the peninsular India, in spite of its massive catchment area, the discharge is not very impressive because of moderate annual average rainfall in the basin. The waste water generation from domestic (both rural and urban) and the industrial sector are the main sources of pollution in the river. Around 17.94 per cent of the BOD load is from industrial sources. BOD is a measure of water pollution based on the organic matter it contains. The organic matter provides food for aerobic bacteria, which require oxygen to be able to bring about the bio-degradation of such pollutants. The greater the volume of organic material and greater the number of bacteria, the greater will the demand for oxygen. Also higher values of COD indicating the higher microbial activities and presence of oxidizable organic matter. Khulbe and Durgapal (1993) reported similar findings.

Materials and Methods

(a) Sampling Sites—Three sampling sites Station A (before Someshwar temple), Station B (Ramkund, Panchavati) and Station C (Tapovan) were selected and samples were collected at the interval of one month for two years, i.e. November 2001 to October 2003. Station B and Station C are receiving polluted water from various sources.

(b) Analysis of Samples — Water samples were analysed in the laboratory following Winkler's idiometric modified azide method (1989) for DO and for BOD, samples were collected in dark bottles and incubated at 20°C for five days and were analysed by volumetric method. COD was determined by dichromate reflexing method suggested by Moore et. al. (1949).

Results and Discussion

Monthly mean values of DO, BOD and COD are given in Table 1.1, 1,2 and are represented graphically in Figures 1.1 to 1.6 respectively. High organic constituents are responsible for a significant portion of oxygen demand (Babu et. al. 1997). The organic matter supports good microbial growth, the microorganisms utilize the oxygen and lead to low values of dissolved oxygen and consequently high values of BOD and COD.

Dissolved Oxygen

DO is the important factor for aquatic life. The maximum DO 8.1 mg/lit was recorded in winter in the month of December at Station 'A' in both the years and minimum DO 2.9 mg/lit and 3.0 mg/lit was recorded in summer in the month of June at Station 'C' in the year 2001-02 and 2002-03 respectively. High values of DO during winter could be due to greater solubility of oxygen in water at lower temperature. The depletion of dissolved oxygen values due to discharge and degradation of wastes at three stations indicated that the Godavari River was polluted.

Biochemical Oxygen Demand

During the present investigation the values of BOD observed were maximum 15.0 mg/lit to 15.7 mg/lit, in summer in the month of May at Station 'C' and minimum 3.5 mg/lit to 3.7 mg/lit in winter in the month of January, at Station 'A' during both the years. The BOD values were found high in summer followed by monsoon and low in the winter, similar observations were made by Qumerunnisa (1985). Low BOD values in winter may be due to suspended solids settle on the bottom. The water becomes clear and the microbial population decreases due to low temperature. The same trend is observed by Varghese et. al. (1992).

Chemical Oxygen Demand

The maximum and minimum values of COD were recorded from 92.4 mg/lit to 40.0 mg/lit at three stations during both the years. The COD values were slightly increasing from pre-monsoon and were highest in monsoon during present investigation. Similar trend was reported by Kudesia et. al. (1986). High COD shows the presence of a huge amount of accumulated organic matter and its incomplete oxidation.

Conclusion

In the present investigation on Godavari River for following parameters like DO, BOD and COD clearly indicates the pollution of river water which is unsuitable for domestic use and drinking purpose. This may be due to release of municipal sewage, domestic wastes and large number of anthropogenic activities may be carried out along the river banks. If the quality of river water is to be maintained in good condition, necessary action should be taken by the authorities to prevent it from further deterioration.

Acknowledgement

The authors are grateful to the coordinator, M.G. Vidyamandir for providing necessary facilities to carry out research work.

References

Khulbe, R.D. and A.. Durgapal (1993): Evaluation of drinking water quality at Bhimtal, Nainital, Uttar Pradesh, *Poll. Res.* 21(2): 109-11.

Kudesia, K.P., S.P. Verma, Singh and Sanjiv (1986) : Pollution studies of drinking water quality of environmentally degraded village Kamalpur (Distt. Meerut) and their remedial measures, *Indian J. Environ. Agric.* 1(1) : 38-44.

Moore, W.A., R.C. Kroner and C.C. Ruchhoft (1949): *Anal. Chem.*, 21: 923.

Qumerunnisa (1985) : Ecobiology of fresh water ciliates, Ph.D. Thesis, Marathwada University, Aurangabad, M.S., India.

Varghese, M., A. Chauhan and L.P. Naik (1992): Hydrobiological studies of a domestically polluted tropical pond 1. Physico-chemical characteristics, *Poll. Res.* 11: 95-100.

Table-1.1 : Monthly Mean values of DO, BOD and COD mg/l from Godavari river water at Nashik during the year November 2001- October 2002

Months	*DO (Stations)*			*BOD (Stations)*			*COD (Stations)*		
	A	**B**	**C**	**A**	**B**	**C**	**A**	**B**	**C**
November 01	7.00	6.80	6.70	4.10	4.20	4.20	59.40	60.40	61.80
December 01	8.10	7.80	7.60	3.90	4.20	4.30	42.20	44.20	44.80
January 02	7.90	7.20	7.10	3.70	3.90	3.80	40.00	42.40	42.90
February 02	6.90	6.40	6.20	4.20	3.90	3.80	54.60	58.00	58.90
March 02	5.90	5.70	5.40	4.50	4.90	5.20	56.90	58.80	59.80
April 02	4.50	4.40	4.10	5.90	6.10	6.30	57.30	62.40	64.20
May 02	3.30	3.40	3.10	14.20	14.70	15.00	68.20	68.00	67.20
June 02	3.20	3.30	2.90	13.70	13.90	14.00	69.30	70.50	72.00
July 02	3.80	3.60	3.60	12.20	12.50	12.90	88.40	89.30	87.40
August 02	4.90	4.60	4.20	10.90	11.50	11.80	88.80	92.40	88.50
September 02	5.80	5.40	5.30	9.20	9.50	9.80	60.30	63.40	67.20
October 02	5.50	5.10	5.00	8.50	8.90	9.50	55.40	58.40	59.80
Average	5.56	5.30	5.10	7.91	8.18	8.38	61.73	64.01	64.54

Table 1.2 :

Monthly Mean values of DO, BOD and COD mg/l from Godavari river water at Nashik during the year November 2002-October 2003

Months	*DO (Stations)*			*BOD (Stations)*			*COD (Stations)*		
	A	B	C	A	B	C	A	B	C
November 02	7.20	6.90	6.70	3.80	3.90	4.10	62.20	62.50	62.90
December 02	8.10	7.80	7.50	3.60	3.80	4.10	44.10	45.30	46.00
January 03	7.80	7.60	7.20	3.50	3.80	3.80	43.00	44.90	45.50
February 03	6.80	6.50	6.30	4.10	3.60	3.60	54.90	56.20	58.50
March 03	6.30	5.80	5.40	4.80	5.40	5.80	56.70	59.50	60.20
April 03	4.90	4.60	4.20	5.80	6.20	6.40	57.90	62.60	63.90
May 03	3.40	3.30	3.00	1530	15.50	15.70	68.50	72.40	70.50
June 03	3.00	2.80	3.00	13.90	14.00	14.20	70.10	72.80	70.80
July 03	3.80	3.70	3.70	12.50	13.00	13.20	88.20	99.40	87.80
August 03	4.80	4.40	4.30	11.00	11.90	12.50	90.00	90.50	88.30
September 03	5.70	5.20	5.20	9.90	10.00	10.20	58.30	60.70	61.60
October 03	5.40	5.30	5.10	9.00	9.30	9.50	56.20	58.70	59.20
Average	5.60	5.32	6.31	8.10	8.36	8.59	62.50	64.62	64.60

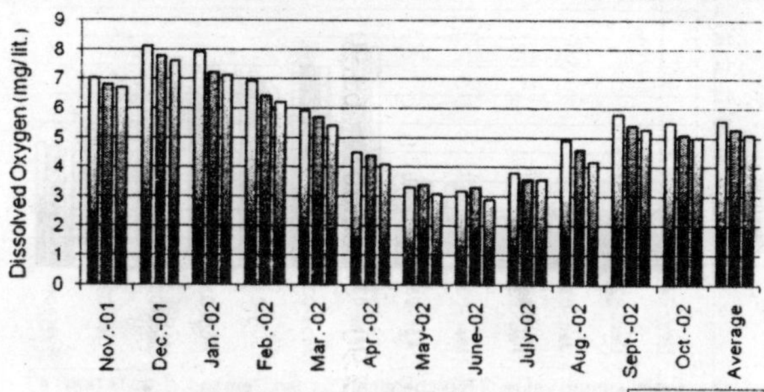

Fig 1.1 : Monthly mean value of Chemical Oxygen Demand COD) (mg/lit.) of water samples at various sampling stations during the monitoring period (Nov. 2002 to Oct. 2003).

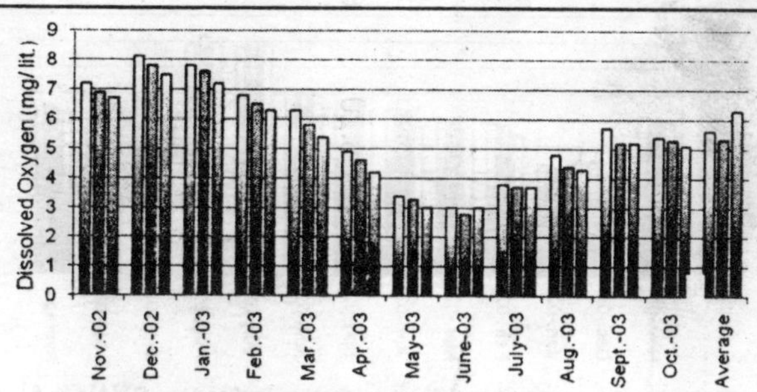

Fig 1.2 Monthly mean value of Dissolved oxygen (mg/lit.) of water samples at various sampling stations during the monitoring period (Nov. 2002 to Oct. 2003)

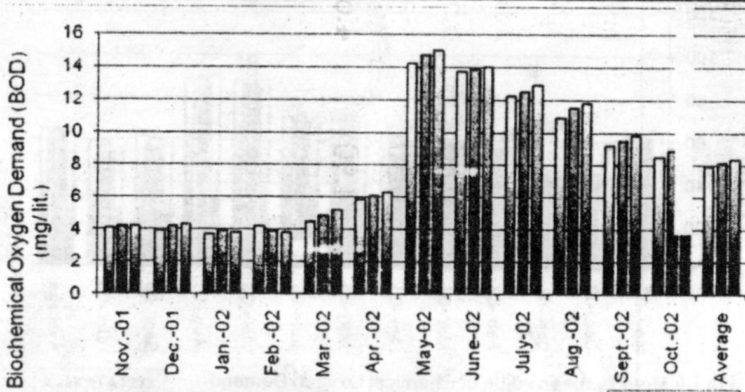

Fig 1.3 Monthly mean value of Biochemical Oxygen Demand (BOD) (mg/lit.) of water samples at various sampling stations during the monitoring period (Nov. 2001 to Oct. 2002).

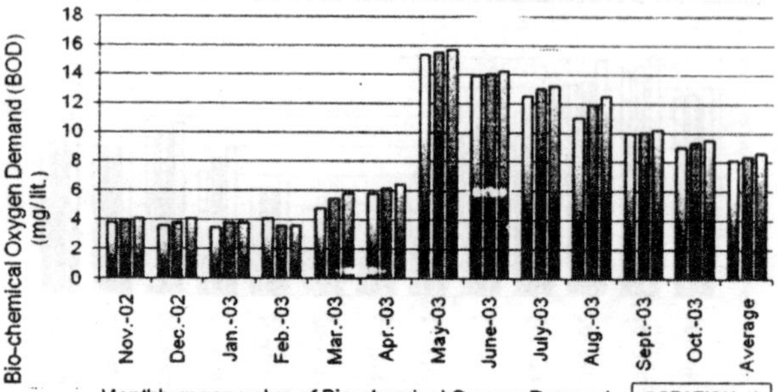

Fig 1.4 Monthly mean value of Bio-chemical Oxygen Demand (BOD) (mg/lit.) of water samples at various sampling stations during the monitoring period (Nov. 2002 to Oct. 2003).

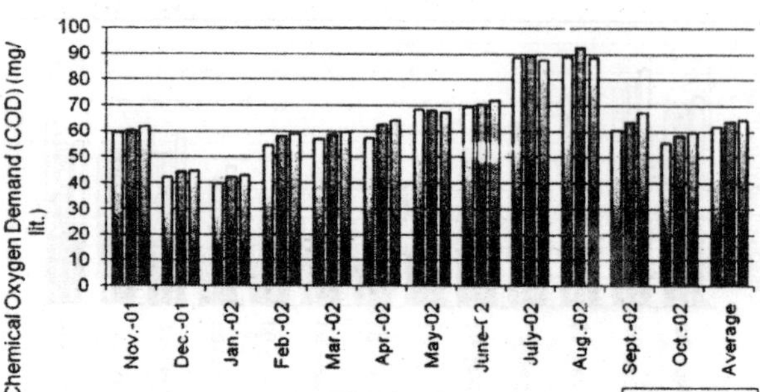

Fig 1.5 Monthly mean value of Chemical Oxygen Demand COD) (mg/lit.) of water samples at various sampling stations during the monitoring period (Nov. 2002 to Oct. 2003).

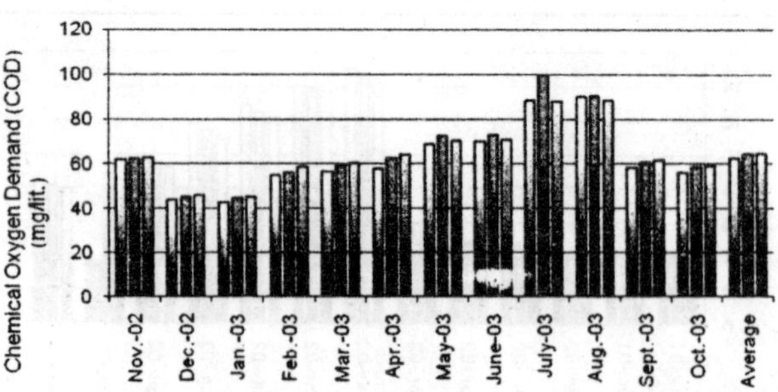

Fig 1.6 Monthly mean value of Chemical Oxygen Demand (COD) (mg/lit.) of water samples at various sampling stations during the monitoring period (Nov. 2002 to Oct. 2003).

2

Study of the Physico-Chemical Properties of the General Soil of Malegaon Tehsil

B.S. Yadav, R.R. Pawar and Alkesh Ahire

ABSTRACT

The Physico-Chemical parameters of field soil of Malegaon tehsil was studied. The bulk density and particle density was highest in several sample. The water holding capacity of soil was high. The highest sodium content was detected in some sample (201). The calcium carbonate percentage was medium. Potassium, which form the essential macronutrient. The content of potassium was fluctuating and moderate. The results of the present investigation are presented in Tables 2.1, 2.2 and Figures 2.1 to 2.8 clearly indicate that there is a lot of variations among the soil samples collected from different fields of Malegaon tehsil.

Keywords : Soil porosity, pycnometer, organisms, water holding capacity.

Introduction

Life on earth depends directly on the living soil and the aquatic ecosystem of the rivers, without fertile soil and the microbial fauna that inhibit its food would not grow, dead things would not decay and nutrients would not be recycled.

Yet the earth's soil are being stripped away rendered sterile and contaminated with toxic chemicals at a rate that cannot be sustained. Most soil areas used for agriculture either lack water or have physical and chemical constraints like steep slopes, easily eroded or poory drained, soil alkalinity and other conditions are toxic to plants. The present experiment was carried out to determine the types and amount of nutrients present in the soil solution because certain proportion of nutrients present in the soil, make the soil fertile.

Material and Methods

In the present experiments soil sample were collected from eight fields of various villages of Malegaon tehsil (Annexure l). The sample collection site is designated as GF1 to GF8. The soil sample were collected from 15 to 20 cm depth and thoroughly mixed on a clean piece of cloth and kept for drying on flat tray. It is important to maintain the identity of each sample at all stages of preparations. Drying causes changes in the physical and chemical characteristics of soil. When soil is air dried stones pieces and macro. organic matter are picked out and weighed. After weighing sample was crushed and sieved. Large lumps are broken up by hand and them the soil is grinded by rolling gently with a wooden roller. After grinding, soil is screened through a 2 mm sieve. The sample is used for analysis. Following parameters were analysed by using the appropriate methods described by APHA, Jalotra et. al. Hesse et. al. Tardan Brown et. al. and Guggenheim.

1. Bulk Density
2. Water contents
3. Porosity
4. Air filled porosity
5. PH
6. Lime potential
7. Conductivity
8. Calcium carbonate
9. Organic matter
10. Sodium and Potassium

Results

The results of the present investigation are presented

in Table 2.1. and 2.2 and Figures 2.1 to 2.8. The result reveals that bulk density was highest in GF5 (1.7333) followed by GF8 (1.1114) and GF 4 (1,1062). The lowest bulk density was recorded in GF2 (0.9102). This may be due to high water content of the soil. Particle density of GF1 shows highest value (2.2317) and lowest at GF3 (1.6155).

The high water holding capacity is good characteristics of soil. The high water content was measured in GF1 (22.88 per cent) and lowest water content was recorded in GF2 (11.62 per cent).

The high porosity is favourable for the microbial growth, water holding and root propagation of plants. The GF1, GF2 and GF3 shows high porosity value 0.5287, 0.4609 and 0.4304 respectively. The low porosity value observed in GF 8 (0.2742). The highest air filled porosity was observed in GF2 (0.2716) and lowest air filled porosity was in GF6 (0.0376) Table-2.11.

The soil sample showed pH between the range of 8 to 8.9 except GF2 (8.01), similarly other parameter were measured and results are shown in Tables 2.1, 2.2 and Figures 2.1 to 2.8. Lime potential was measured highest in GF6 (7.37) followed by GF8 (7.36) and GF5 (7.32). The lowest lime potential measured in GF 1 which is 7.02. The electrical conductivity was measured highest in GF8 (0.44) and lowest value observed in GF4 and GF5 (0.161). The highest value of organic carbon and organic matter was recorded in GF1 (1.10 and 1.896) followed by GF6 (1.07, and 1.845) and lowest in GF8 (0.98 and 1.689). The GF6 (4.25) GFl (4.287) and GF7 (4.33) were observed non-calcareous while GF2 (5.49), and GF8 (5.4084) were medium calcarious in nature. The highest sodium content was observed in GF6 (201) followed by GF7 (183.8) and GF2 (132.4), the lowest sodium was shown in GF4 (33.41) and GF3 (40.5). The present results showed the highest potassium in GF8 (978.8) than the other sample. (Table 2.1, 2.2 and Figures 2.1 to 2.8).

Discussion

Soil is the naturally occurring, unconsolidated or loose

covering of broken rock particles and decaying organic matter on the surface of the earth, capable of supporting life.

All soils initially comes from rocks, this is termed the parent materials. Soil is the main rooting medium for agriculture plants and long-term cultivation affects the fertility of the soil, and makes the causes of low yield of crops, therefore, time to time analysis of sc l fertility parameters must be carried out to get the satisfa :tory yield.

The results of the present investigation pr ·vide us not only physical but also chemical parameters th: ough which we can understand actual requirement of agricultural fields. Aggelides and Londra (2000) reported the fluctuation in porosity, bulk density and pH. In the present study it was observed that the soil samples are in the way of becoming alkaline in nature. Srikanth et. al. (2000) reported that the alkalinity of the soil decrease by adding the enriched form waste, compost enriched with weeds and cow dung slurry. The results of the present probe shows normal electrical conductivity, the highest electrical conductivity (0.44) recorded at GE8 and lowest at GF5 (0.16 l). Similar results were reported by Hortenstine and Rathwell (1972).

Organic carbon and organic matter level was found fluctuating in all the soil samples and its level is enough hence there is no need to supply externally biofertilizers. Hadas, et. al. (2004) observed similar results and further reported that composted municipal waste increase the soil organic matter. Similar observation were made by Hoffman (1983).

Overall results are in the range of common values of parameters being analysed but not completely under the actual recommended values still it not so harsh that it can damage fertility of soil of each respective fields.

Acknowledgement

Authors are very much thankful to the principal, MSG College Malegaon for providing the laboratory facilities.

References

APHA (1989), *American public Health Association, Standard methods for the Examination of Water and Waste,* 19th edn. Washington DC.

Brawn, R. B. (2003), *Soil Texture, Fact Sheet SL 29,* University of Elorida, Institute of Food and Agricultural sciences, Retrieved on 2007. Buol S.W. F. D. Hole and R.J. Mc Cracken (1973).*Soil genesis and Classification 1st. Ames.* lA Iowa state University Press.

Guggenheim, Stephen and R.T. Martin (1995) Definition of clay and clay mineral, *Journal report of the AIPEA nomenclature and capital nomenclature committees,clay and clay minerals,* 43, 255-56.

Hadas, A., M. Agassi, H. Zheviele, L. Kautsky, G.J. Levy, B. Fizik and M. Gotessman (2004). Mulching with composted municipal solid wastes in the central Negev, Israel in *Effect on available nitrogen and phoshorus on organic matter in soil tillage res.* 78, 115-28.

Hoffmann, G. (1983), Recycling of municipal waste compost in vineyard soil In: *Studies about humas* ed. Novak. B. Transaction of VII international symposium Vol. 1 Abstracts praque, p. 69.

Hartenstine, C.C. and D.E. Rathwell (1972). The use of Municipal compost in reclaimation of phosphate mining sand tailings, J. Env. Qual, 1: 415-18.

Hesse, P. R. Chap 1. Origin and nature of soil, Chap. 2 Preparation of soil sample for analysis, Chap. 3 Soil reaction, pH value and lime potentials, *A Textbook of Soil Chemical Analysis.*

Jalitra, S. K. Romes, Khera and B.S. Ghyman, Chap. 2 State properties of soil 16-61, *Methods in Soil Physics* (Narsoa Publishing House).

Shrikant, K. (1997), Enrichment of compost and its effect on soil properties and crop growth, Ph.D. Thesis, U.A.S. Bangalore India.

Tardon H.L.S. *Analysis of soil for available major nutrients, methods of analysis of soil, plants, water and fertiliser.*

Table 2.1 : Physical Properties of Different PGF soil Samples

Sample	Db	Dp	SWC%	F	Fa	pH	L.P.	EC	%CaCO3
GF1	1.0519	2.2317	22.88	0.5287	0.1363	8.6	7.02	0.225	4.267
GF2	0.91021	1.6886	11.62	0.4609	0.2716	8.01	7.29	0.211	5.49
GF3	1.09745	1.6155	18.776	0.4304	0.0862	8.4	7.2	0.162	5.37
GF4	1.1062	1.7316	16.45	0.3612	0.04809	8.4	7.25	0.161	5.4084
GF5	1.13338	1.8057	15.7316	0.3723	0.06406	8.96	7.32	0.161	5.2784
GF6	1.1004	1.7618	18.112	0.3754	0.0376	8.84	7.37	0.197	4.25
GF7	1.0879	1.8041	17.6747	0.397	0.06956	8.9	7.28	0.221	4.334
GF8	1.11145	1.80831	11.85	0.2742	0.03859	8.7	7.36	0.44	5.4084

Graph 2.1 : Comparison of Dp and Db

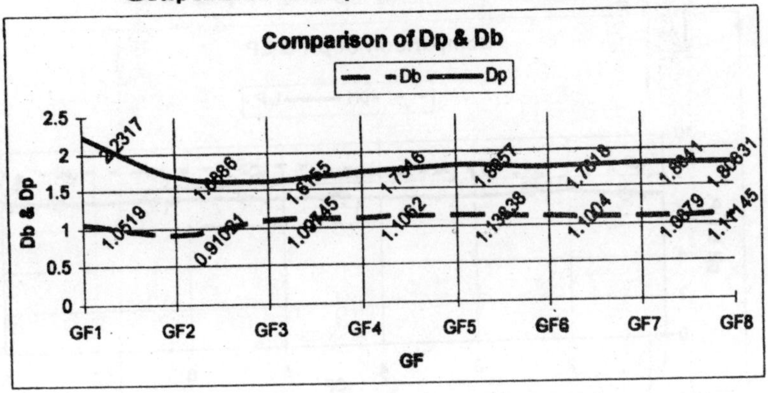

Graph 2.2 : Comparison of SWC

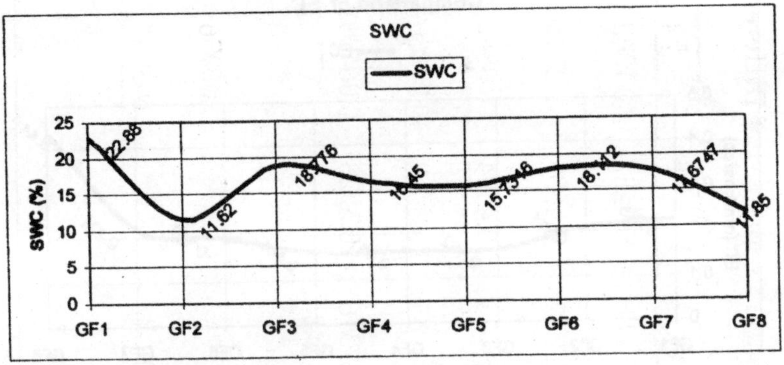

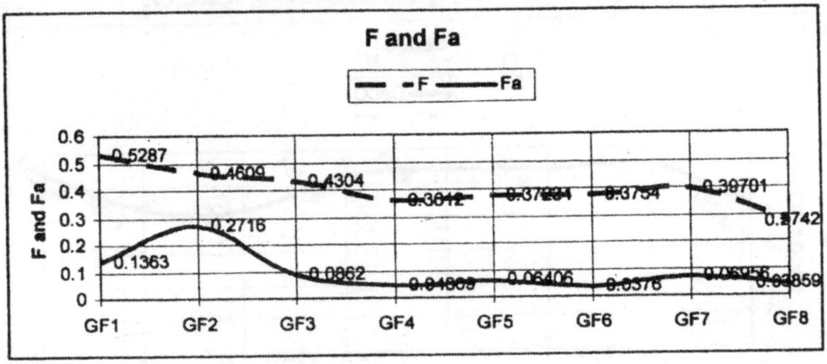

Graph 2.4 : Comparison of pH & LP

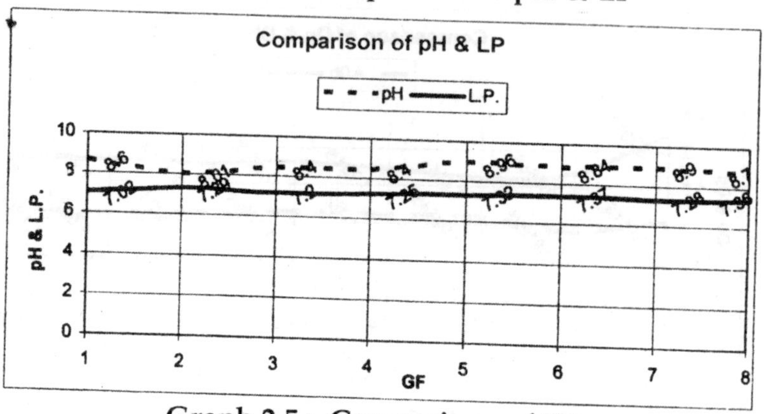

Graph 2.5 : Comparison of EC

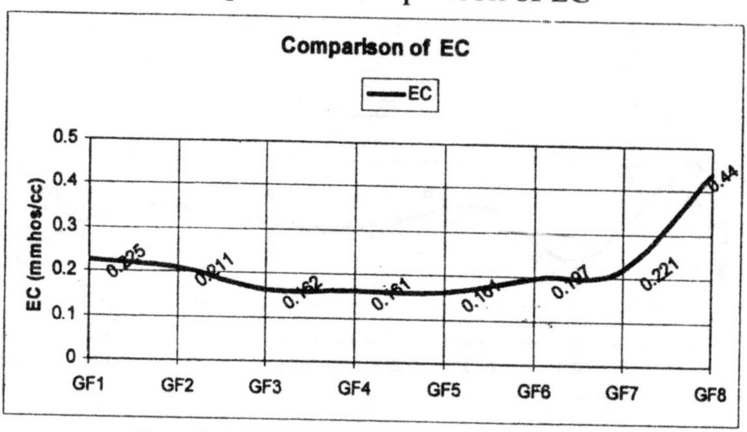

Graph 2.6 : Comparison of % $CaCO_3$

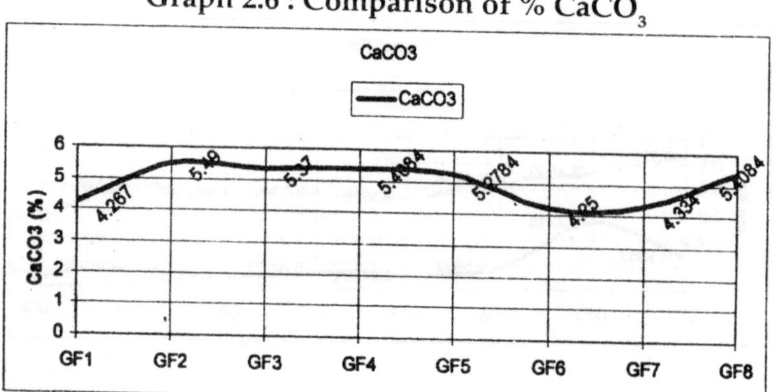

Table 2.2 : Chemical properties of different PGF Soil samples

Sample	*OC*	*OM*	*Na*	*K*
GF1	1.1	1.896	113	465
GF2	1.01	1.741	132.4	400
GF3	1.05	1.81	40.5	665
GF4	1.03	1.776	33.41	202.5
GF5	1.02	1.758	60.18	157.5
GF6	1.07	1.845	201	787.5
GF7	1.06	1.827	183.8	540
GF8	0.98	1.689	77.06	978.8

Graph 2.7 : Comparison of OC and OM

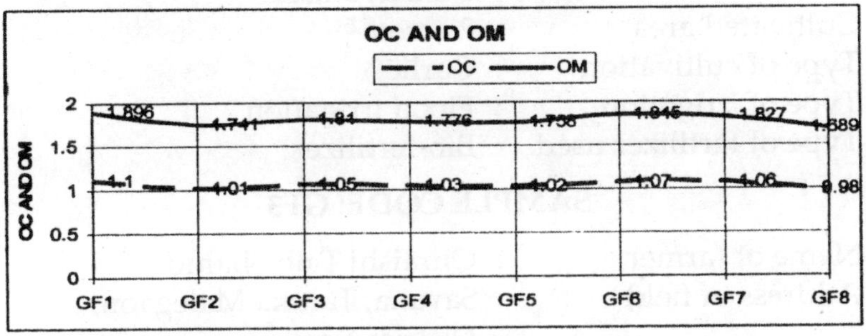

Graph 2.8 : Comparison of Na and K

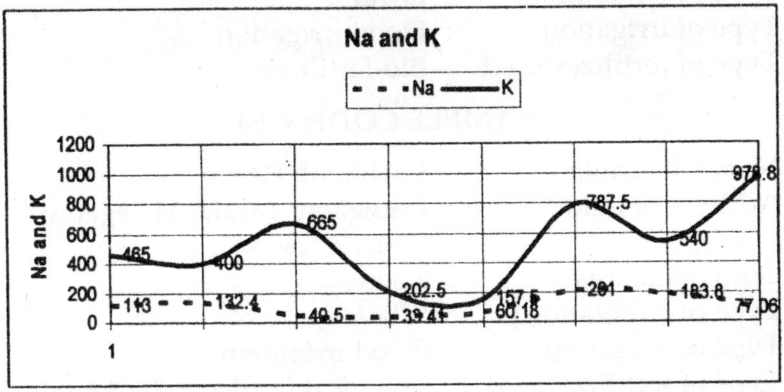

ANNEXUE-1

SAMPLE CODE: GF1

Name of farmer	:	Quraishi Taib Shahid
Address of field	:	Sayana, Taluka Malegaon, District Nasik
Cultivated area	:	2 acres
Type of cultivation	:	Onion
Type of irrigation	:	Flood irrigation
Type of fertilizer used	:	Bio- fertilizer

SAMPLE CODE: GF2

Name of farmer	:	Quraishi Taib Shahid
Address of field	:	Sayana, Taluka Malegaon, District Nasik
Cultivated area	:	2 acres
Type of cultivation	:	Garlic
Type of irrigation	:	Flood irrigation
Type of fertilizer used	:	Bio-fertilizer

SAMPLE CODE: GF3

Name of farmer	:	Quraishi Taib Shahid
Address of field	:	Sayana, Taluka Malegaon, District Nasik
Cultivated area	:	2 acres
Type of cultivation	:	Jawar
Type of irrigation	:	Flood irrigation
Type of fertilizer used	:	Bio-fertilizer

SAMPLE CODE: GF4

Name of farmer	:	Kashinath Pawar
Address of field	:	Daregaon, Taluka Malegaon, District Nasik
Cultivated area	:	2 acres
Type of cultivation	:	Wheat
Type of irrigation	:	Flood irrigation
Type of fertilizer used	:	Chemical fertilizer (18: 18: 00)

SAMPLE CODE: GF5

Name of farmer : Ashok Patil
Address of field : Malda, Taluka Malegaon, District Nasik
Cultivated area : 3 acres
Type of cultivation : Onion
Type of irrigation : Flood irrigation
Type of fertilizer used : Urea

SAMPLE CODE: GF6

Name of farmer : Kashinath Pawar
Address of field : Dyana, Taluka Malegaon, District Nasik.
Cultivated area : 2 acres
Type of cultivation : Garlic
Type of irrigation : Flood irrigation
Type of fertilizer used : Bio-fertilizer

SAMPLE CODE: GF7

Name of farmer : Subhash Nikam
Address of field : Syoegaon near power sub-station Taluka Malegaon, District Nasik
Cultivated area : 2 acres
Type of cultivation : Mixed plantation
Type of irrigation : Drip irrigation
Type of fertilizer used : Bio-fertilizer

SAMPLE CODE: GF8

Name of farmer : Subhash Nikam
Address of field : Syoegaon near power-sub station Taluka Malegaon, District Nasik
Cultivated area : 3 acres
Type of cultivation : Mixed plantation
Type of Irrigation : Drip irrigation
Type of fertilizer used : Bio-fertilizer

ANNEXUE 2
Percentage value of N, P and K present in different chemicalfertilizers

Name of Fertilizers	*% Nitrogen*	*% Phosphate Oxide*	*% Potassium Oxide*
Nitrogen containing Fertilizers			
1. Ammonium Sulphate	20.6	-------------	-------------
2. Calcium Ammonium Nitrate	25	-------------	-------------
3. Ammonium Sulphate nitrate	26	-------------	-------------
4. Urea	46	-------------	-------------
Phosphate containing Fertilizers			
1. Single Super Phosphate	-------------	16	-------------
2. Tripel Super Phosphate.	-------------	42.5	-------------
3. Dicalcium Phosphate	-------------	26	-------------
Potassium containing Fertilizers			
1. Murate of Potash	-------------	-------------	58
2. Sulphates of Potash	-------------	-------------	48
1. Nitro Phosphate	20	20	0
2. Nitro Phosphate (with Potash)	15	15	15
3. Mono Ammonium Phosphate	11	55	0
4. Di Ammonium Phosphate	18	43	0
Mixed Fertilizers	34	17	0
	16	16	16
	10	36	10
	12	61	0
	11	19	19
	20	10	10

ANNEXUE 3

Percentage value of N, P, and K present in different bio-fertilizers.

Sources of biofertilizers	*(%)Nitrogen*	*(%)Phosphate Oxide*	*(%)Potassium Oxide*
Compost	0.80	0.65	1.00
Goat dung	0.60	0.50	0.70
Poultry	3.03	2.36	1.40
Cow dung	0.56	0.35	0.78
Human excreta	1.30	1.10	0.35
Groundnut cake	7.10	1.40	1.30
Cotton seed cake	6.40	2.80	2.50
Caster seed cake	4.50	1.70	0.70
Bone crush	3.50	21.50	0.00
Fish wastage	4.10	0.90	0.30

3

Limnological Studies of Algae of Khandesh Area of Maharashtra

Ziya Ansari
S.N. Nandan

ABSTRACT

Algal and water samples were collected at monthly interval from 4 stations of Mausam River passing through Malegaon city. The seasonal fluctuations of Physico-chemical parameters such as pH, DO, alkalinity, hardness and TDS were observed for 4 stations of river at monthly interval. The different algal taxa were encountered in present investigation. The correlation of physico-chemical parameters and algal periodicity was made in present study. the pollution-toterant genera of 4 groups of algae were recorded in present study analysis of water samples made according to APHA.

Keywords : Algae, Physico-chemical parameters, Pollution.

Introduction

The important sources of pollution in Mausam River include discharge of raw sewage at different bathing ghats, the disposal of dead animals and discharge of feacal matter, etc. A great reduction in the number and kinds of algae has been frequently recorded by various workers following the exposure of algae to gross pollutional stress.

During the last few decades algae have been used as fascinating tool in the assessment of water quality. Since the ecosystem in such situation is exposed to altered environmental conditions, it often happens that the sensitive species disappear leaving behind the tolerant forms. The organisms able to resist the adverse or stressed habitat are sometimes designated as indicators of pollution.

In present investigation attempts have been made to correlate the algal community structure with the physico-chemical characteristics of polluted Mausam River.

Material and Methods

The malegaon city is situated along the bank of the Mausam River. The algal and water samples were collected at monthly intervals from 4 stations of Mausam river. The physico-chemical analysis of water samples made according to APHA (1985). The algal population of algal samples were estimated in the laboratory with Haemocytometer. Algal taxa encountered in present investigation were identified with standard monographs.

Results and Discussion

In present study, Chlorophyta was represented by algal forms, viz., *Spirogyra, Ulothrix, Chlorella, Stigeoclonium, Oedogonium, Cosmarium,* and *Scenedesmus*. A marked difference in the occurrence of alga species was recorded at different sampling sites. Another group of algae encountered Cyanophyta which comprises mainly the species of *Microcystis, Oscillatoria, Phormidium, Lyngbya* and *Spirulina*. The maxium growth of this group was recorded during summer characterized by high temperature which played important role in the periodicity of this group of algae as emphasized by many workers (Hutchinson, 1967; Rai and Kumar, 1976; Rai, 1978; Nandan and Patel, 1981).

The diatom flora of the Mausam River was found to be much richer than other algal groups. The flora of diatoms was represented by species of *Fragilaria, Nitzchia, Cymbella, Narivulla, Synedra, Gomphonema, Gyrosigma, Pinnularia,* etc. In diatoms, Nitzschia was commonest and most dominant form as it was recorded consistently in larger numbers. The numbers of *Euglinineae* were presented by *Euglena, Phacus* and *Trechelomonas*. In present study, blue-green and diatoms were dominant as compared to other groups of algae.

The ranges of physico-chemical parameters are given in Table-3.1. The major physico-chemical and biological features of Mausam river were the alteration between the polluted and unpolluted regions. In present study the physico-chemical data reveals that concentration of hardness and chloride was greater as compared to other parameters. The concentration of nitrate was more when compared to other parameters. The concentration of nitrate and phosphate was increased from station 1 to onwards in down stream. It is observed that polluted water do not tend to follow the general rule on observation similar to that of Rao (1971) and Khan et. al. (1978).

Table 3.1 : Physico-chemical parameters of 4 stations of Mausam river

	Stations			
Physico-Chemical Parameters	*S1*	*S2*	*S3*	*S4*
Water Temperature	15-32°C	15-32°C	15-32°C	15-32°C
PH	5.09-8.73	5.18-8.65	5.42-8.72	5.72-8.69
Dissolved Oxygen	10.3-18.44	10.15-18.58	3.46-18.15	2.65-15.5
Free CO_2	0-1.5	0-0	0-17.6	0-13.64
Carbonate	0-60	0-30	0-80	0-36
Bicarbonate	40-220	40-220	20-170	30-180
Total Alkalinilty	60-245	20-210	20-170	30-180
Chloride	22.72-660	21.3-1200	17-1420	5.6-1086
Hardness	44-650	162-920	250-950	240-876
Calcium	9.61-61.72	13.6-101.8	11.2-84.16	8.8-89.73
Total dissolved Solids	78-179	98-756	68-675	85.977
Phosphate	0.03-1.82	0.05-1.19	0.09-1.92	0.37-1.94
Nitrate	0.03-1.5	0.8-1.52	0.13-1.54	0.14-1.57

All parameters are expressed in mg/l except water temperature and pH.

Air temperature and water temperature did not show

any remarkable difference. The pH of all stations range from 5.42-8.91 (Table-3.1). In present study the higher concentration of chlorides increased from station 1 to onwards in down stream. Seasonal fluctuation of physico-chemical parameters has observed for 3 seasons.

A considerable work has been done on the effects of temperature and hight in relation to diatoms and it has postulated by many workers that diatoms have critical range of temperature for their growth and development (Hutchinson, 1967). It is tallied with present investigation. Diatoms were also found in high polluted environments of all stations confirming the earlier results (Rai and Kumar, 1977; Nandan and Patel, 1985). The pollution tolerant genera of algae were recorded.

The chemical analysis revealed that all stations are polluted and the concentration of chloride, phosphate, nitrate and total dissolved solids was increased from station 1 to onwards confirming the results of Nandan & Patel (1985).

Acknowledgement

We are grateful to Principal B.M. Patil for providing the facility for investigation.

References

A.P.H.A. : *Standard methods for the estimation of water and waste water,* 16th ed., American Pub. HHL. Asso, New York, 1985.

Hutchinson, G.E. : *A treatise on limnology,* Vol. II Introduction to lake biology and the limnoplanktons, John Wiley and Sons, Inc. 1967.

Khan, A.A., A.G. Siddiqui : A Musharraf and Tazia Hameed, Physico-chemical and biological characters of pond Chau-Tal, *J. Zool Res.* (Aligarh) 1978 2:1-13.

Nandan, S.N. & R.J. Patel : Ecological studies on algal flora of Vishwamitri River, Baroda, Gujarat, *J.P.I. Nature* 1984 1(1) : 17-22

Nandan, S.N., R.J. Patel: Pollution studies of Vishwamitri river flowing through Baroda, *Bioviganam* 1985: 11(2) 209-210.

Rai, L.C. : Ecological Studies on algal communities of river Ganges at Varanasi, *Ind. J. Ecol.*, 1978 5:1-6.

Rai, L.C. & H.D. Kumar : Systematic and ecological studies on algae of some habitats polluted with fertilizer factory effluent near Sahupuri, Varanasi India, *Nova Hedwigia*, 1976, 27: 825-8 15.

Rai, L.C. & H.D. Kumar: Studies on the seasonal variation in the algal communities of pond polluted with fertilizer factory effluent, *Indian J. Ecol.* 1977, 4(2):124-31.

Rao, V.S. : An ecological study of 3 fresh water ponds of Hyderabad, India I. *The environment. Hydrabiologia,* 1971 38 (2) : 351-72.

4

Quality of Groundwater of Bore Wells in Malegaon Camp, Nashik, Maharashtra

Yuvraj Sonawane

ABSTRACT

The investigation on the physico-chemical characteristics of the bore well water samples in the Malegaon Camp carried out from April 2008 to March 2009 show that the parameters are within the range prescribed by BIS, WHO and ISI standards for drinking water. The bore well water is safe for domestic needs after proper treatment.

Keywords : Physico-chemical parameters, groundwater, quality, Malegaon Camp.

Introduction

Ground water is one of the earth's most important resources. All life on earth depends on water. It is a universal solvent and is a major constituent of all living organisms, is a well-known phenomenon (Gupta & Gupta, 1997). The increase demand of population growth, urbanization, industrialization and irrigation stress in ground water resources (Hasan, 2003). The work in this area of ground water analysis was carried out by Manoranjini et. al. (1998), Rawat (2003), Patil (2003), Dhembare & Pondhe (1998),

Johnson & Kauser (2004), Pawar (2006), Mishra *(2005)*, Kumar (2004), Tripathi & Pandey (2009), Chatterjee (2008), Mahajan (2008), Agarwal (2009), etc. In many parts of India, especially the rural parts, people face scarcity of potable water. So also there is no proper distribution system which compels the people to fetch water for bore wells. Normally the bore wells go as deep as 200 feet or more. In many cases, people complain about health problems. It has prompted to undertake this research work.

Materials and Methods

Malegaon is the second largest city located 20°32' north and 74°35' east in the Nashik district of Maharashtra State. The study was carried out in bore wells water samples in Malegaon Camp. The distance between the two stations is approximately 1 km. For physico-chemical analyses — 10 sampling stations were selected and sampling was done from April 2008 to March 2009 in 2 liter plastic bottles and small tubes, respectively. For the analysis of water, procedure given in standard methods (APHA, 1989), Trivedy & Goel (1986) and Trivedy et. al. (1987).

Results and Discussion

(a) *Water temperature :* The minimum (18°C) was recorded at M1 station in January 2009 and the highest (27.5°C) at station M9 in May 2008. It was higher in summer, lower in winter and medium in rainy season.

(b) *pH* : Minimum pH (6.84) was observed in July 2008 at station M_6, while maximum (8.62) at station M2 in January 2009. pH in the range of 6.5 to 8.5 indicated the suitability of water for all purposes.

(c) *DO and BOD :* Maximum DO level (5.73 mg/l) was at stations M_5 in November 2008, while minimum (3.65 mg/l) at station M_8 in August 2008. Minimum BOD level (0.92 mg/l) was noted at station M_5 in

September 2008, while maximum (3.6 mg/l) at stations M_2 and M_6 in April and May 2008. Patil & Dongare (2006) and Sonawane & Saler (2008) also pointed out that DO values showed decreased, while BOD showed an increased trend.

(d) *Total Alkalinity* : Maximum value of total alkalinity (443 mg/l)) was noted in August 2008 at station M10, while minimum (215 mg/l) in December 2008 at station M_5. The presence of carbonates, bicarbonates and hydroxide are the major cause of alkalinity in natural water (Jain, et. al., 1997).

(e) *Conductivity:* The electrical conductivity value varies within the range of 0.564 to 5.103 μmhos indicating low mineralization in this region. It generally affects the taste and acceptance of the water as potable. Piska et. al. (2004). The present result was within the range of potable water standards.

(f) *TDS* : Minimum TDS value (0.423 ppm) was observed in December 2008 at station M1, while maximum (3.418 ppm) in March 2009 at station M10. Todd (1980), when TDS of fresh water is less than 1000 ppm, it can be used safely for drinking purpose.

(g) *Hardness* : Total hardness varied between 166 to 468 mg/l. According to Kannan (1991) water with hardness values more than 180 mg/l is very hard. The standard unit for the total hardness specified is 300 mg/l and is considered potable; but beyond this limit it produces gastrointestinal irritation (ICMR, 1975). In the present investigation the values of hardness were variable, i.e. this water can be included in hard to very hard category.

(h) *Calcium* : Maximum values of calcium (72.38 mg/l) was noted in September 2008 at station M_8, while minimum (25.09 mg/l) in January 2009 at stations M_1 and M_7. The standard unit for the calcium specified is 75 mg/l. The results indicate that ground water is suitable for domestic purpose.

(i) *Magnesium* : Minimum value of magnesium (24.51 mg/l) was recorded in February 2009 at station M_2, while maximum (87.22 mg/l) in November 2008 at station M_8.

(j) *Chloride* : The chlorides varied from 65.32 to 275.9 mg/l and the values of chloride was found to be within the permissible limit as prescribed by BIS.

(k) *Acidity:* Minimum amount of acidity (2.5 mg/l) was noted in December 2008 at station M1, while maximum (120.75 mg/l) in September 2008 at station M_8.

(l) *Nitrate*: Nitrate of the ground water was noted between 0.1 to 0.58 mg/l. The concentration of nitrate above 45 mg/l in drinking water may prove harmful to human health (Jain et. al., 1997). The present value was within the range of potable water standard.

(m) *Phosphate* : The phosphate varied in the range of 0.01 to 0.18 mg/l. Overall phosphate concentration was low in the water.

On the basis of above physico-chemical parameters it may be concluded that the water of bore wells is within range prescribed by WHO, ISI and BIS standard for drinking water. The present results showed that water is safe for domestic needs after proper treatment.

Acknowledgement

The author gratefully acknowledges the financial assistance given by U.G.C. Western Region Office, Pune towards minor research project. I am also thankful to Shri Prashantdada Hiray, General Secretary, M.G. Vidyamandir, Malegaon Camp and Dr. B.S. Jagdale, Principal, M.S.G. College, Malegaon Camp for providing the library and laboratory facilities. Thanks are due to Dr. D.C. Pendse, Ex-Head, Department of Botany for encouragement throughout this work.

Table 4.1 : Physico-chemical parameters at 10 stations from April 2008 to March 2009

Parameters	Water Temp (°C)									
Stations	M1	M2	M3	M4	M5	M6	M7	M8	M9	M10
April	26.0	26.5	26.4	27.0	26.4	27.0	25.8	26.9	27.0	26.5
May	26.5	26.6	26.0	27.2	26.5	27.1	26.6	27.3	27.5	26.9
June	26.3	26.3	25.0	26.8	26.1	26.7	26.1	27.0	27.3	26.0
July	26.0	26.1	25.0	26.0	26.0	26.2	26.0	26.8	26.5	25.8
August	25.0	25.1	24.1	25.2	25.2	25.0	24.8	25.0	26.0	24.0
September	24.0	24.1	23.0	25.0	24.0	24.8	24.0	24.2	25.1	23.0
October	23.8	23.0	22.0	24.5	23.1	24.3	23.6	23.5	23.0	22.5
November	22.2	22.0	21.8	24.1	22.2	24.0	21.0	22.3	21.0	22.0
December	18.1	20.0	20.5	23.6	20.0	23.2	18.5	21.0	20.2	21.6
January	18.0	18.1	18.5	23.0	18.3	22.0	18.2	20.5	20.0	21.0
February	19.3	19.0	19.2	24.5	19.8	23.5	19.6	21.6	21.3	22.5
March	20.0	20.3	20.8	25.6	20.8	24.6	20.8	22.8	22.6	23.5

Parameters	pH									
Stations	M1	M2	M3	M4	M5	M6	M7	M8	M9	M10
April	7.35	7.48	7.58	7.52	7.68	7.52	7.68	7.43	7.25	7.36
May	7.18	7.15	7.42	7.41	7.51	7.39	7.52	7.32	7.12	7.18
June	7.08	7.12	7.32	7.32	7.42	7.21	7.33	7.15	7.01	7.09
July	6.92	7.02	7.18	7.08	7.15	6.84	7.18	6.78	6.89	7.01
August	7.19	7.22	7.35	6.91	7.05	7.15	7.02	7.08	7.35	7.15
September	7.53	7.52	7.91	7.15	7.18	7.35	7.15	7.38	7.62	7.35
October	7.84	7.78	8.02	7.29	7.44	7.51	7.48	7.57	7.78	7.48
November	8.12	7.99	8.15	7.42	7.68	7.69	7.72	7.68	7.99	7.62
December	8.23	8.34	8.32	8.00	8.12	8.08	8.12	7.98	8.12	7.92
January	**8.35**	**8.62**	**8.43**	**8.15**	**8.25**	**8.22**	**8.32**	**8.12**	**8.42**	**8.15**
February	8.22	8.53	8.37	8.06	8.12	8.14	8.21	8.05	8.27	7.98
March	7.97	8.41	8.25	7.94	7.93	7.89	8.08	7.88	8.12	8.82

Parameters	DO (mg/l)									
Stations	M1	M2	M3	M4	M5	M6	M7	M8	M9	M10
April	4.59	4.99	4.93	4.79	4.65	4.39	4.93	4.19	4.39	4.99
May	4.19	4.99	4.65	4.65	4.53	4.25	4.85	4.05	3.85	4.85
June	4.05	4.85	4.79	4.53	4.19	4.13	4.59	3.85	3.79	4.65
July	3.93	4.13	4.45	4.25	3.99	3.79	4.25	3.73	3.73	4.45
August	**3.79**	**3.73**	**4.05**	**4.25**	**3.85**	**3.73**	**4.05**	**3.65**	**3.85**	**4.33**
September	4.53	4.39	4.45	4.59	4.05	4.05	4.53	3.93	4.13	4.53
October	4.99	4.53	4.93	5.39	4.59	4.65	4.99	4.39	4.45	4.65
November	**5.13**	**5.45**	**5.52**	**5.45**	**5.73**	**4.73**	**5.13**	**4.79**	**4.73**	**4.85**
December	5.19	5.65	5.45	5.33	5.39	4.53	5.19	4.93	4.79	4.93
January	4.99	5.13	4.99	5.19	5.13	4.45	**5.25**	**4.85**	**4.19**	5.05
February	4.12	4.92	5.84	5.04	4.98	4.98	5.12	4.78	4.38	4.92
March	3.84	4.52	4.5	4.84	4.65	4.78	4.7	4.44	4.04	4.58

Parameters					BOD (mg/l)					
Stations	M1	M2	M3	M4	M5	M6	M7	M8	M9	M10
April	1.4	3.6	2.68	2.74	2.4	2.34	1.4	2.26	2.2	2.6
May	2.0	2.54	1.66	2.58	2.46	3.6	1.32	2.52	2.06	2.72
June	2.2	1.63	2.74	2.6	2.34	2.08	1.34	2.46	**1.4**	3.00
July	1.94	2.34	2.6	2.12	2.54	2.34	1.6	2.34	2.00	2.66
August	1.2	1.67	2.46	2.52	2.00	2.14	2.4	2.32	2.06	2.54
September	1.8	1.86	2.52	1.54	0.92	1.66	1.88	1.74	2.08	2.14
October	1.67	2.00	2.48	2.06	1.26	1.8	2.2	1.6	2.06	2.46
November	1.8	2.42	2.58	2.4	1.4	1.74	2.34	1.4	2.54	2.32
December	1.87	2.52	1.8	1.94	1.86	1.88	2.4	2.08	2.4	2.68
January	2.4	2.6	2.34	2.6	1.88	1.86	2.63	2.32	1.8	2.46
February	2.26	2.14	2.46	2.74	2.28	2.34	2.13	2.2	2.28	2.08
March	2.06	2.42	2.4	2.66	2.66	2.86	2.86	2.74	2.34	2.66

Parameters				Total Alkalinity (mg/l)						
Stations	M1	M2	M3	M4	M5	M6	M7	M8	M9	M10
April	363.5	360	370	371.5	356.5	363.5	350.0	365.0	368.5	375.0
May	380.0	381.5	386.5	381.5	368.5	375.0	363.5	373.5	375.0	380.0
June	383.5	385	396.5	388.5	373.5	390.0	378.5	410.0	388.0	393.0
July	386.5	390	403.5	413.5	376.5	398.5	380.0	411.5	435.0	440.0
August	388.5	393.5	406.5	416.5	388.5	425.0	388.5	416.5	436.5	443.0
September	350.0	303.5	376.5	365.0	315.0	353.5	338.5	375.0	378.0	375.0
October	340.0	293.5	323.5	358.5	283.5	340.0	311.5	350.0	360.0	351.5
November	303.5	251.5	253.5	308.5	233.5	306.5	286.5	333.5	305.0	296.5
December	293.5	241.5	250	296.5	215.0	300.0	261.5	301.5	286.5	278.0
January	306.5	245.0	261.5	300	225.0	298.5	271.5	293.5	290.0	293.0
February	311.5	251.5	268.5	308.5	226.5	306.5	280.0	301.5	300.0	303.0
March	321.5	261.5	275	313.5	235.0	315.0	291.5	308.5	310.0	316.5

Parameters	Conductivity (umhos/cm.)									
Stations	M1	M2	M3	M4	M5	M6	M7	M8	M9	M10
April	1.07	1.05	1.05	1.22	1.43	1.16	1.07	0.99	1.34	1.24
May	1.11	1.15	1.19	1.29	1.37	1.22	1.15	1.81	1.12	1.35
June	1.16	1.24	1.36	1.43	1.52	1.32	1.30	1.97	1.26	1.56
July	1.788	2.06	3.355	2.823	1.989	2.17	2.472	3.196	2.23	3.096
August	2.471	2.717	3.914	3.836	2.983	3.742	3.661	4.681	3.354	4.624
September	2.766	3.357	4.429	4.681	3.458	4.336	4.324	5.103	3.970	5.102
October	2.658	3.1	4.063	4.735	3.148	4.425	4.18	4.369	4.178	4.08
November	1.001	1.027	1.713	2.941	1.916	2.938	2.182	3.21	2.773	2.044
December	0.837	0.712	1.041	1.482	0.796	1.099	1.14	1.324	1.049	1.336
January	0.626	0.564	1.077	1.034	0.56	1.03	1.094	1.045	0.971	1.043
February	0.848	0.981	2.082	2.029	1.053	2.194	1.816	2.162	1.959	2.309
March	1.682	2.111	2.835	3.127	1.803	3.174	2.610	3.718	2.910	3.344

Parameters	TDS (ppm)									
Stations	M1	M2	M3	M4	M5	M6	M7	M8	M9	M10
April	0.597	0.622	0.676	0.892	0.721	0.854	0.638	**0.936**	**0.909**	**0.758**
May	0.613	0.692	0.702	0.998	0.764	0.974	0.78	1.109	1.017	0.809
June	0.621	0.779	0.793	1.014	0.803	1.121	0.841	1.225	1.118	0.884
July	0.657	0.849	0.884	1.113	0.892	1.295	0.929	2.031	2.015	0.931
August	0.742	0.897	0.977	1.289	0.969	1.467	1.123	2.113	2.073	1.048
September	0.623	0.761	0.813	1.002	0.821	1.126	1.233	2.124	2.126	1.13
October	0.558	0.694	0.704	0.88	0.745	0.947	1.125	2.741	2.12	1.008
November	0.491	0.573	0.619	0.731	0.602	0.817	0.993	2.005	1.909	0.947
December	0.423	0.592	0.597	0.795	0.563	0.799	0.852	1.146	1.118	0.893
January	0.482	0.672	0.627	0.84	0.621	0.864	0.715	0.986	0.974	0.908
February	0.975	1.006	1.127	1.317	1.222	1.522	1.017	1.922	2.005	2.117
March	**1.885**	**2.127**	**2.417**	**2.137**	**2.585**	**2.62**	**2.221**	**2.952**	**3.218**	**3.418**

Parameters	Hardness (mg/l)									
Stations	M1	M2	M3	M4	M5	M6	M7	M8	M9	M10
April	188	186.6	194	260	210	256	198.6	262.6	224.6	224.6
May	214	200.6	198.6	286.6	216.6	266.6	209.4	288.6	227.4	230.6
June	230	210	204.6	301.4	234.6	288.6	216	306.6	244.6	238
July	241.4	226	220	320.6	256	304.6	228	314	266.6	256
August	260.6	292	301.4	414.6	370	280	280	400	344.6	325.4
September	330	321.4	340	432	306.6	370	354.6	440	360	380
October	297.4	334	354	444	280	410	360.6	466.6	401.4	400
November	220	293.4	334	430.6	276.6	394.6	320	468	380	380
December	181.4	228	224	366	226.6	316.6	233.4	382.6	306.6	280
January	184	166	193.4	300	180	258	197.4	273.4	253.4	213.4
February	191.4	174.6	198.6	304	186.6	262	204	278.6	258	219.4
March	198	181.4	204	307.4	190	266	211.4	285.4	264.6	228

Parameters	Calcium (mg/l)									
Stations	M1	M2	M3	M4	M5	M6	M7	M8	M9	M10
April	26.45	26.21	31.50	41.44	25.09	42.72	27.25	39.51	43.52	40.08
May	26.69	26.45	31.82	41.92	25.41	43.52	27.81	39.83	44.33	40.32
June	27.25	27.01	32.30	42.24	25.41	44.32	28.05	40.88	45.13	40.88
July	30.46	27.81	33.90	44.32	25.41	46.25	28.85	42.48	47.29	42.24
August	35.02	31.82	41.12	46.73	36.31	47.05	32.06	44.09	46.25	45.69
September	52.66	46.25	52.67	67.89	43.29	64.36	39.28	72.38	60.92	50.26
October	52.90	45.93	53.70	68.69	44.32	66.29	40.32	71.34	65.17	51.06
November	30.46	28.61	32.06	40.64	30.46	40.32	25.89	44.08	39.51	37.91
December	25.65	26.21	31.5	32.30	28.61	32.06	25.65	33.90	31.82	32.62
January	25.09	25.89	31.02	29.66	25.41	29.89	25.09	29.89	28.86	31.82
February	26.21	29.65	33.66	31.82	28.05	31.82	26.69	32.06	31.26	32.30
March	29.09	32.3	35.51	33.66	31.02	35.02	28.61	35.27	34.7	34.46

Parameters	Magnesium (mg/l)									
Stations	M1	M2	M3	M4	M5	M6	M7	M8	M9	M10
April	29.72	29.53	28.11	38.15	35.91	36.40	31.82	39.96	28.26	30.36
May	35.91	32.79	29.04	44.34	37.32	38.49	34.11	46.09	28.45	31.67
June	39.47	34.74	30.21	47.75	41.71	43.36	35.57	49.85	32.16	33.13
July	41.61	38.15	32.98	51.16	46.92	46.09	38.00	50.68	36.20	36.69
August	42.19	51.80	48.43	72.60	44.14	61.54	48.72	70.66	55.84	51.50
September	48.39	50.19	50.82	63.98	48.39	51.01	62.51	63.20	51.99	62.03
October	40.29	53.45	53.60	66.41	45.80	59.59	63.34	70.31	58.18	66.41
November	35.08	54.09	61.89	80.20	48.88	71.63	62.22	87.22	68.56	69.53
December	28.60	39.61	35.42	69.53	37.81	57.30	41.27	72.60	55.35	48.39
January	29.58	24.85	28.26	55.06	28.40	44.68	32.84	48.43	44.19	32.64
February	30.69	24.51	27.92	54.72	28.40	44.49	33.48	48.39	43.86	33.81
March	30.55	24.56	28.11	54.41	27.43	43.51	34.11	48.09	43.37	34.59

Parameters	Chloride (mg/l)									
Stations	M1	M2	M3	M4	M5	M6	M7	M8	M9	M10
April	98.4	96.13	104.65	123.54	97.55	107.92	104.08	125.39	118.29	109.77
May	115.44	101.81	126.8	143.42	115.44	129.22	106.92	146.69	129.64	137.31
June	131.06	136.74	137.74	204.05	129.22	147.68	127.8	204.48	148.67	131.06
July	148.1	156.62	151.94	248.5	145.83	188.86	143.42	242.39	191.27	244.67
August	158.61	175.08	172.81	269.8	163.3	199.22	161.45	275.48	201.21	275.05
September	163.3	177.5	173.67	270.79	164.72	200.22	162.3	275.05	202.63	275.9
October	112.8	116.87	127.8	183.6	106.5	149.52	137.74	146.69	116.01	139.16
November	89.89	81.93	95.14	112.6	87.04	96.13	93.72	104.09	103.23	98.97
December	73.84	67.73	77.1	98.4	69.58	86.19	79.09	93.29	84.2	87.04
January	72.42	65.32	73.41	96.99	66.31	85.62	77.67	91.87	83.35	85.62
February	76.25	68.58	77.1	103.23	71.42	88.04	86.62	96.56	87.61	89.46
March	78.52	71	82.78	106.4	75.68	92.29	91.87	99.82	92.72	94.14

Parameters	Acidity (mg/l)									
Stations	M1	M2	M3	M4	M5	M6	M7	M8	M9	M10
April	PA=44.25	PA=50.75	PA=48.25	PA=65.00	PA=45.00	PA=59.25	PA=51.75	PA=62.5	PA=51.75	PA=55.00
May	PA=53.25	PA=59.25	PA=53.25	PA=71.75	PA=54.25	PA=66.75	PA=61.75	PA=68.25	PA=63.25	PA=64.25
June	PA=78.28	PA=75.75	PA=73.25	PA=85.75	PA=75.0	PA=99.25	PA=78.25	PA=102.5	PA=82.5	PA=89.25
July	PA=89.25	PA=80.0	PA=75.75	PA=110.75	PA=82.5	PA=109.25	PA=84.25	PA=115.75	PA=105.0	PA=92.5
August	PA=91.75	PA=84.25	PA=86.75	PA=113.25	PA=85.0	PA=111.75	PA=88.25	PA=118.25	PA=106.75	PA=95.0
September	PA=92.5	PA=86.75	PA=87.5	PA=115.75	PA=86.75	PA=113.25	PA=90.0	PA=120.75	PA=109.25	PA=95.75
October	PA=26.75	PA=22.5	PA=32.5	PA=39.25	PA=24.25	PA=28.25	PA=27.5	PA=35.75	PA=36.75	PA=33.25
November	PA=5.0	PA=3.25	PA=7.5	PA=15.75	PA=5.75	PA=15	PA=5.75	PA=16.75	PA=15.0	PA=15.0
December	PA=2.5	PA=4.25	PA=6.75	PA=8.25	PA=3.25	PA=5.75	PA=5.0	PA=10.0	PA=4.25	PA=11.75
January	PA=2.25	PA=5.75	PA=9.25	PA=10.75	PA=4.25	PA=8.25	PA=7.5	PA=12.5	PA=6.75	PA=13.25
February	PA=19	PA=6.75	PA=10.75	PA=11.75	PA=5.75	PA=9.25	PA=8.25	PA=13.25	PA=10	PA=15
March	PA=19.75	PA=8.25	PA=11.75	PA=12.5	PA=5.75	PA=10.75	PA=9.25	PA=14.25	PA=9.23	PA=16.75

Parameters	Nitrate (mg/l)									
Stations	M1	M2	M3	M4	M5	M6	M7	M8	M9	M10
April	0.16	0.10	0.10	0.33	0.10	0.25	0.10	0.16	0.16	0.16
May	0.16	0.16	0.10	0.40	0.10	0.33	0.10	0.25	0.25	0.25
June	0.25	0.25	0.16	0.43	0.10	0.33	0.16	0.33	0.43	0.25
July	0.33	0.25	0.16	0.43	0.16	0.43	0.16	0.43	0.43	0.43
August	0.33	0.33	0.25	0.50	0.16	0.43	0.25	0.43	0.50	0.43
September	0.40	0.33	0.25	0.50	0.16	0.50	0.25	0.58	0.58	0.50
October	0.25	0.16	0.16	0.43	0.10	0.33	0.16	0.50	0.43	0.33
November	0.16	0.16	0.10	0.25	0.10	0.25	0.10	0.43	0.33	0.25
December	0.10	0.16	0.10	0.16	0.08	0.16	0.10	0.33	0.16	0.16
January	0.10	0.08	0.08	0.10	0.08	0.10	0.08	0.16	0.10	0.16
February	0.16	0.16	0.10	0.16	0.10	0.16	0.10	0.25	0.16	0.25
March	0.25	0.25	0.16	0.25	0.16	0.25	0.16	0.33	0.25	0.33

Parameters	Phosphate (mg/l)									
Stations	M1	M2	M3	M4	M5	M6	M7	M8	M9	M10
April	0.03	0.04	0.04	0.10	0.03	0.03	0.04	0.08	0.06	0.04
May	0.04	0.04	0.06	0.12	0.04	0.04	0.04	0.08	0.08	0.06
June	0.04	0.04	0.06	0.15	0.04	0.04	0.06	0.10	0.10	0.06
July	0.04	0.06	0.06	0.18	0.06	0.06	0.06	0.15	0.12	0.08
August	0.06	0.06	0.08	0.18	0.08	0.10	0.06	0.15	0.15	0.08
September	0.06	0.10	0.10	0.18	0.18	0.12	0.08	0.18	0.18	0.10
October	0.04	0.08	0.08	0.15	0.08	0.10	0.06	0.15	0.15	0.08
November	0.03	0.03	0.03	0.06	0.03	0.04	0.03	0.10	0.06	0.04
December	0.01	0.03	0.03	0.04	0.01	0.03	0.01	0.03	0.03	0.03
January	0.01	0.01	0.03	0.03	0.01	0.03	0.01	0.03	0.03	0.03
February	0.03	0.03	0.06	0.06	0.03	0.04	0.03	0.06	0.06	0.04
March	0.04	0.04	0.08	0.08	0.06	0.06	0.06	0.08	0.08	0.06

All parameters are expessed in mg/l, except pH, water temperature °C
Ta= Total alkalinity, Conductivity µmhos/cm.,
PA = Phenolphthalein acidity.

References

Agarwal, S.K., *Water Pollution*, APH Publishing Co., New Delhi, 2009.

APHA, AWWA, WPFC, *Standard Methods for the Examination of Water and Wastewater*, 17th ed., American Public Health Association, Washington D.C., 1989.

BIS, *Indian Standard of Drinking Water Specification*. Bureau of Indian Standards, New Delhi, 1993.

Chatterjee, S.N., *Water Resources, Conservation and Management*, Atlantic Publishers & Distributors (P) Ltd., New Delhi, 2008.

Dhembare, A.J. and G.M. Pondhe, Physico-chemical characteristics and quality of ground water in the Chinchpur area, Maharashtra State. *J.Aqua.Biol.*, 1998, 13, 52-53.

Gupta, S. and M. Gupta, *Water and its Management: Water and Basic Environmental Technology*, Amol Publication Pvt. Ltd., New Delhi, 1997.

Hasan, S.Z., Ground Water exploitation and hydrogeological reseach : A chronology of historical developments. 2003, *J Curr.Sc.*, 2, 299-304.

ICMR, *Manual of standards of Quality of Drinking Water Supplies*, Special Report Series no. 44, 1975.

ISI. Indian Standard Specification for Drinking Water, ISI 10500.

Jain, C.K., K.K.S. Bhatia and T. Vijay, Ground water quality in coastal region of Andhra Pradesh. *Indian J. Env.Hlth.*, 1997, 39, 182-92.

Johnson, Mary Esther Cynthia and Rana Kauser, Chemical and microbial quality of different types of drinking water of Hyderabad, HI-Tech city, A.P., India, *J Aqua.Biol.*, 2004, 19, 93-97.

Kumar, Arvind, *Water Pollution*, APH Publishing Co., New Delhi, 2004.

Mahajan, Gautam,. *Evaluation and Development of Ground Water*, APH Publishing Co., New Delhi, 2008.

Manoranjini, Kiran C. and Mary Esther Cynthia Johnson. Quality analysis of bottled and its source water, a study on water quality assessment. *J. Aqua. Biol.*, 1998, 60-61.

Mishra, S.R., *Advances in Limnology*. Daya Publishing House, Delhi, 2005.

Narayana, Shenoy. K. and K.N. Lokesh, Quality of ground water of bore wells in M.I.T., Campus, Manipal, Kamataka, *Indian J. Environ Hlth.*, 1999, 41, 144-48.

Patil, Dyneshwari and Meena Dongare, Seasonal variations in dissolved oxygen and biochemical oxygen demand of same lentic water bodies of Koihapur city, Maharashtra. *Geobios,* 2006, 33, 70-72.

Patil, S.S., J.M. Pitwari and R.G. Auti, Ground water quality from Pravara area, District Ahmednagar, Maharashtra, *J. Aqua. Biol.*, 2003, 18, 85-86.

Pawar, Anusha C., S. Jithender Kumar Naik, Naresh Jadhav, V. Vasundhara, Devi and Smita C. Pawar, Physio-chemical study of ground water samples from Nacharam industrial area, Hyderabad, Andhra Pradesh, India, J. *Aqua. Biol.*, 2006, 21, 118-20.

Piska, R.S., P.Y. Swamy and R. Parvathy, Heavy metal pollution and its toxic effect on ground water quality of Jeedimeetla IDA, Hyderabad, *Indian J. Env. Prot.*, 2004, 24, 177-81.

Rawat, Mamta, Comparison of quality of ground water resources of arid region in Rajasthan, India, *J. Aqua. Biol.*, 2003, 18, 61-63.

Sonawane, Yuvraj, D. and R.S. Saler, Hydrobiological study of a village pond near Malegaon, Maharashtra, *Geobios,* 2008, 35, 157-60.

Sonawane, Yuvraj D. and R.S. Saler, Physico-chemical characteristics of water reservoir near Malegaon, Maharashtra. *Geobios.* 2008, 35, 25-28.

Todd, D.K., *Groundwater, Hydrology,* 2nd edn. John Wiley and Sons, New York, 1980.

Tripathi, A.K. and S.N. Pandey, *Water Pollution.* APH. Publishing Co., New Delhi, 2009.

Trivedy, R.K. and P.K. Goel, *Chemical and Biological Methods for Water Pollution Studies.* Environmental Publications, Karad, 1986.

Trivedy, R.K., P.K. Goel and C.L. Trisal, *Practical methods in Ecology and Environmental Science,* Environmental Publications, Karad, India, 1987.

WHO, *Guidelines for Drinking Water,* Vol.2. CBS Publisher & Distributors, New Delhi, 1991.

WHO, *International Standards for Drinking Water.* World Health Organization Tech. Report, 1984.

5

Eutrophication of Mausam River of Malegaon City by Algal Communities

S.N. Nandan
Ziya Ansari

ABSTRACT

Four different stations of Mausam river flowing through Malegaon city were selected for collection of water and algal samples at monthly intervals for September 1992 to August 1993. The identification of algal taxa of 4 groups of algae, viz., Bacillariophyceae, Cyanophyceae, Chlorophyceae and Euglenineae was made. Nygaard's trophic state indices are used for assessing the water quality and eutrophication of Mausam river. In present study all stations of river were eutrophic in nature. All indices of eutrophication hold good for assessing the water quality of 4 stations of Mausam river of Malegaon city of Maharashtra.

Keywords : Eutrophication, Algae, Mosam River.

Introduction

Eutrophication means the increase of growth of algae and other aquatic plants as a consequence of increased input of mineral nutrients. It is directly or indirectly connected with pollution because the mineral nutrients evolve from organic waste water by microbial decomposition. Weber (1907) introduced first the concept of eutrophication. Nygaard (1949) introduced five indices for evaluation of eutrophication. According to him certain algal groups were indicative of levels of nutrient enrichment. In present paper Nygaard's indices are used to get meaningful evaluation of eutrophication in Mausam river of Malegaon city of Maharashtra.

Materials and Methods

Algal and water samples were collected at monthly intervals for 2 years from 4 stations of Mausam river which is situated north-south direction of Malegaon city of Maharashtra. Algal samples were preserved in 4 per cent formalin for further study. For the study of eutropication Nygaard's (1949) trophic state indices were used for determination of trophic state of 4 stations of Mausam river of Malegaon city in present study.

Results and Discussion

In the present study five Nygaard's indices were calculated (Table-5.1). The lower number of desmids indicates the higher eutrophic nature of water. In present study total algal taxa of desmids were less at all 4 stations of Mausam river. Trophic state indices of Myxophycean and Chlorophycean are always inversely proportional to the number of desmids while the diatoms and euglenophyte indices are quite independent. The myxophycean quotient was showing eutrophic nature at all 4 stations. It was the highest at station 4 (Table-5.1). The Chlorophyceae index was eutrophic at all 4 stations in present study as the number of desmids were less as compared to Chlorococcales at all 4 stations of river. The diatoms quotient was showing oligographic nature at all 4 stations as the number of pinnate diatoms were more than that of centric diatoms. It was the least sensitive. The index of euglenophyte showed eutrophic nature at all stations except station 1 because the station 1 showed less numbers of euglenoids as compared to other stations. The compound quotient is much useful and sensitive for determining the eutrophication because it was assessed by considering the all types groups of algae at 4 stations of river (Table-5.1). The index of compound was greater at all 3 stations as compared to station-1 in present study (Table-5.1).

The most eutrophic form like *Microcystis* was recorded in present study. Also the most eutrophic forms of diatoms and euglenoids were recorded at Station-3 and 4 of Mausam River in present study. The number of euglenoids was less at Station1 as compared to other groups of algae as it was less organically polluted.

Throughout the investigation the dominance of blue-greens at all stations of Mausam river indicating highly eutrophic nature of river as agreed with views of earlier workers (Hasler, 1947; Nandan and Patel, 1985). Also the occurrence of bloom of *Mycrocystis* during summer has been recorded in present study indicating eutrophic nature of river as it was confirmed with earlier observations of Rawson (1956).

Today the most of workers preferred a synthetic approach of combining individual with population quotients as probably the best method of assessing eutrophication. In present study the trophic state indices according to Nygaard (1949) hold good for assessing the eutrophication of 4 stations of Mausam river as it was agreed with results of Gunale & Balakrishnan (1981) and Nandan and Patel (1985).

Acknowledgement

The authors are grateful to principal B.M. Patil, S.S.V.P.S's L.K. Dr. P.R. Ghogrey Science College, Dhule for providing facilities for present investigation.

References

Gunale, V.R. and Balkrishnan, M.S. 1981. Biomonitoring of Eutrophication in the Pavana, Mula and Mutha rivers flowing through Poona. *Indian J.Environ. Hlth.* 23(4): 316-22.

Hasler, A.D. 1947. Eutrophication of lakes by dcmestic drainage. *Ecology* 28: 383-95.

Nygaard G. 1949. Hydrobiological studies in some Danish ponds and lakes-Il. The quotient hypothesis and some new or little known phytoplankton organism. *Bio.Sci.* 7: 1-293.

Nandan, S.N. and Patel R.J. 1985. Eutrophication in Vishwamitri river flowing through Baroda city. *Geobios* 11:60-62.

Rawson D.S. 1956. Algal indicators of trophic lakes types. *Limnol. Oceanogr.* 1: 15-25.

Weber C.A. 1907. Aufbau and vegetation der Moore Norddeut Schlands. *Beibl.Bot.Johrb.* 90: 19-54.

Table 5.1 : Trophic state indices at 4 stations of Mausam river according to Nygaard (1949).

Index	Quotient	Stations			
		S1	S2	S3	S4
Myxophycean	Myxophyceae / Desmideae	3.66	3.16	3.37	3.37
Chlorophycean	Chlorococcales / Desmideae	1.00	1.50	1.25	1.25
Diatoms	Centric diatoms / Pennate diatoms	0.078	0.078	0.071	0.065
Euglenophyte	*Euglenoohyta* / Myxophycee+ Chlorophyceae	0.14	0.44	0.59	0.60
Compound	Myxophyceae + Chlorococales+ Centric Diatoms + Euglenophyta / Desmideae	6.00	8.83	7.87	8.50

6

Hydrobiology of a Waterbody Near Malegaon

Yogesh Shastri
D.G. Jadhav

ABSTRACT

Physico-chemical parameters and the algae of percolation tank of Pimpalgaon near Malegaon Dist. Nasik have been studied from January to December 1997. The studies shows that the water tank is classified as oligotrophic on the basis of water quality criteria. This paper also reports algal diversity. Total 92 algal taxa belonging to 4 classes, i.e. cyanophyceae, alanophycial, chlorophyceae, Bacillariophyceae and Euglenophyceae were recorded.

Keywords : physico-chemical parameters, algae.

Introduction

Hydrobilogical investigations are carried out on man made lakes in India Abbasi et. al.,1996, Madan Mohan Rao et. al.,1996 and Bahura 1998. Similarly number of studies have been conducted on algal diversity from different regions of Maharashtra (Sarode & Kamat 1984). However no such studies are reported from this region. Therefore present work was undertaken.

Malegaon, a taluka headquarter, is the 2nd largest city

located at 20° 32′ N & 74° 35′ E in the Nashik District of Maharashtra State. The percolation task is situated at Pimpalgaon 15 km away from the city.

Materials and Methods

Water samples were collected from the three stations (P1, P2 and P3) of percolation tank for physico-chemical analysis, at monthly intervals between 8 to 10 a.m. in acid washed two litre capacity plastic bottles. The physico-chemical parameters were analysed as per the procedure given in (A.P.H.A.1985).

Algal samples were also collected from all the stations of percolation tank. These samples were collected in small plastic tubes and preserved in 4 per cent formalin for taxonomic identification. Algae were identified by using standard literature and manuscript. This investigation was carried out from January to December 1997

Results and Discussion

The values of physico-chemical parameters are given in Table 6.1. Following is the list of species identified:

A. CYANOPHYCEAE
Aphanocapsa biformis A. Br.
Merismopedia punctata Meyen
Spirulina meneghiniana Zanard ex Gomont
S. subtilissima Kutz ex Gomont
Oscillatoria subbrevis Schmidle
O tambi Woronichin
Phormidium bohneri Schmidle
P. anomala Rao, C.B.
P. ambiguum Gomont Var. major Lemmermann
Anabaena Variabilis Kuetz. Ex Born. Et Flah.

B. CHLOROPIIYCEAE
Chlamydomonas ehrenbergii Gorosh

C. Pseudopertyi Pasch
B. mucicola Schmidle
Trebouxia humicola (Treboux) West et Fritsch
Pediastrum Boryanum (Turp) Mengh.
P. duplex var gracilimum West & West
P simplex Meyen
P. simplex v. duodenarium (Bail)
P. tetras Ehr. Ralfs.
Tetraedron trilobulatum (Reinsch) Hansg.
T. Minimum (A. Braun) Hansg.
Coelastrum microporum Naegeli
Scenedesmus obliqus (Turp.) Kuetz.
S. quadricauda v. longispina (Chodat) G.M. Smith.
S. quadricauda v. quadrispina (Chodat) G.M.Smith
Cladophora glomerata (L) Kutz.
Oedogonium maharashtransis Kamat
Spirogyra biformis Jao
S. Parvula (Trans.) Czurda.
S. rhizopus Jao.
£ schmidtii W & G. S. West
S. granulata Jao
Closterium cynthia De Not
C. cynthia V. Jenneri (Ralfs.) Krieg
C. leibleinii Kuetz.
C. *Venus* Kuetz
Cosmarium angulosum Breb.
C. *circulare Reinsch*
C. *impressulum Elfv.*
C. *Nymannianum Groenb.*
C. *polygonum(Naeg.)Arch.*
C. *subspeciosum* v.Validus Nordst.
C. *subtumidum* Nordst
Euastrum spinulosum Delp

C. BACILLARIOPHYCEAE

Synedra acus Kuetz. V. acula (Kuetz) V.H.

S. ulna (*Nitz.*) Ehr.f.stauro-destituta Pant
S. ulna (Nitz.) Ehr. V. amphirhynchus
Coconeis placentula Ehr
Achnanthes coaretata Breb. V. elliptica krasske
A. breviceps Agardh V. intermedia (Kuetz.) Cleve
B. *inflata (Kuetz.) Grun. F. elatoides* Cholnosky
Mastogloia recta Hustedt V. puichella Voigl
M. recta Hustedt V. nagpurensis V. nov.
Gyrosigma acuminatum (Kuetz.) Rabh.
G. distortum (W. Smith) Cleve V. parkeri Harrison.
G. Kutzingii (Grun.) Cleve
G. maharashtrensis Sp.nov.
Pleurosigma angulatum (Quekett) W. Smith f. Chemburiana
P. elongatum W. Smith V. Karianum (Grun.) Cleve
P. Salinarum Grun.
P. abscurum W. Smith
Caloneis permagna (Bail) Cleve
Navicula cuspisdata Kuetz.
N. cuspidata Kuetz.f. brevirostata Gandhi
N. cuspidata Kuetz. Vambigua (Ehr.) Cleve
N. cuspidata Kuetz. V ambigua (Ehr.) Cleve f. Diminuta A.Cl.
N. gotlandica Grun.
N. halophila (Grun.) Clevef. subcapitata Ostrup
N. pygmea Kuetz.
N. rhyncocephala Kuetz. V. *amphiceros V.* H.
N. viridula Kuetz.
N. viriduloides Gandhi V. *lanceolata* Gandhi
Pinnularia brebissonii (Kuetz.) Cleve
Cymbella lata Grun. V. nagpurensis V. Nov.
C. *tumida* (Breb.) V.H.
C. *tumidula* Grun
C. *tugida* (Greg.) Cleve
C. *ventricosa* Kuetz.V. *arcuata* Skv.
C. *vidarbhensis* sp.nov.
Gomphonema augur Ehr.
G. *constrictum her. V. capitata* (Ehr) Cleve

Nitizschia apiculata (Greg) Grun
N. heufleriana Grun
N. obtusa W. Smith.
N. regula Hustedt V. fennica A.Cl.

D. EUGLENOPHYCEAE

Euglena flava Dang
E. Viridis Ehr
Phacus acuminatus Stokes
P. longicaudatus (Ehr) Luj
P. anamolus Fritsch et Rich.
P. peteloti Lefevre
P. raciborskii Drezepolskii
Trachelomonas ovalis Daddy.

In present study in all 92 algal taxa belonging to four algal groups were recorded. These are Cyanophyceae, Chlorophyceae, Bacillariophyceae and Euglenophyceae.

Seven genera and twelve species of Cyanophycea noted during the present study. In percolation tank March, April and may repersented the period of more Cyanophycean population.

Tucker and Loyd (1984) stated that moderately high temperature supports the growth of Cyanophyceae. Profound Cyanophyceae growth was also recorded during winter months. In present study luxurient growth of Cyanophyceae has been recorded in summer season. During summer season there was high day temperature.

Philipose (1959) emphasized that natural factors like alkalinity, nitrates and phosphates are responsible for luxurient growth of Cyanophyceae Vyas and Kumar, 1968 attributed abundance of Cyanophyceae to higher values of pH, temperature, phosphate, nitrate and relative dissolved oxygen.

Chlorophyceae is represented by Volvocales, Chlorococcales and Desmids. Volvocales members observed in rainy reason from July-September was 29°C & pH was high, similar results were obtained by Shashikant and Anand, 1978. Desmids are represented by *Closterium, Cosmarium and Euastrum.* In present investigation more desmids were observed. Desmids are tolerent to varying chemical and environmental conditions Barhate, 1985 recorded Desmids tolerent to varying conditions of environment.

Table 6.1 : Water quality of percolation tank.

Parameter	Station		
	P-1	P-2	P-3
Temperature°C	14-26	14-26	14-26
pH	6.25-9.03	6.1-10.8	6.20-9.50
D.O.	24.14-24.16	2.073-24.16	2.013-24.18
Free CO_2	00-94.60	00-114.4	00.00-145.2
Alkalinity	20-45	20-45	20-50
Calcium	12.02-39.27	4.008-52.1	4.008-43.28
Magnesium	25.65-129.8	25.26-131.4	21.64-121.68
Phosphate	00-0.28	00-0.29	00-0.23
Nitrate	0.01-0.29	0.02-0.5	0.02-0.42
TDS	0.147-0.168	0.154-0.537	0.155-0.643

The maximum number of diatoms were observed from November to January and in April. Mannikya Reddy (1984), also observed similar results in the river Tungabhadra. Nandan and Patel, (1984) pointed out that diatoms were occurred when water was rich in nitrates and phosphates. In the present investigation it is observed that temperature is the most important factor in controlling the periodicity of diatoms than any other factor.

Percolation tank showed maximum population of Euglenoids during October, November and December. However Kaushik et. al. (1991) have observed more growth of Euglenoids during winter in a pond when water is comparatively clean with sufficient amount of dissolved oxygen. According to Mannikya Reddy (1984), lower pH is responsible for Euglenoid growth. The Euglenoids in percolation tank grow well when pH was 9.09.

Conclusion

On the basis of present study following conclusion can be drawn. Percolation tank studies reveals that the water tank is classified as obligotrophic on the basis of water quality criteria. Percolation tank is rich in algal diversity. In all 92 algal taxa belong to four groups were observed. Temperature is the main factor which controls the periodicity of Cyanophyceae and Diatoms than any other factors. Desmids are tolerent to varying chemical and environmental conditions. Euglenoids grow luxuriently at pH 9.09.

Acknowledgement

Authors are thankful to The General secretary, M.G.Vidyamandir and Principal, S.P.H. College, Malegaon Camp, Dist. Nasik for granting permission & providing necessary laboratory facilities during investigation.

References

Abbasi, S.A., K.S. Bhatiya, A.V.M. Kunhi ad R. S. Soni : Studies on the limnology of the Kuttadi lake (North Kerala). *Eco.Env.Cons.*, 2, 17-27 (1996).

*Standard methods for examination of water and waste water*18th Edition, American Public Health Association, Washington, 1985.

Bahura, C.K. : A study of physico-chemical characteristic of a highly eutrophic temple tank, Bikaner, *J. Aqua. Biol.*, 13, 47-5 1 (1998).

Barhate, V. P. Studies on the algal flora of Vidarbha and Khandesh Maharashtra, Ph.D. Thesis, nagpur University, Nagpur, 1985.

Kaushik, S. M.S. Agarekar and D.N. Saksena, On the planktonic algae of sewage fed Vivek nagar pond at Gwalior. *Poll. Res.*10 (1) : 25-32, 1991.

Madan Mohan Rao, A., V. Narendra Rao and S.K. Mahmood : Assessment of water quality and pollution of Narsingi pond. *Eco.Env. Cons.*, 2, 45-49 (1996).

Manikya Reddy, P. Ecological studies in the river Tungabhadra (A.P.) with special reference to the effect of paper mill Effluents on the river. Ph. D. Thesis. Osmania University, Hyderabad (A.P.) India, 1984.

Nandan, S. N. and R.J. Patel, Ecological studies on algal flora of Vishwamitri river, Baroda (Gujrat), *Indian Plant and Nature* 1(l):17-22; 1984.

Philipose, M.T. Fresh water phytoplankton of Inland fisheries. *Proc.Symp. Algology*, ICAR, New Delhi, 272-91, 1959.

Sarode, P. T. and N. D. kamat : *Fresh water diatoms of Maharashtra*, Saikripa Prakashan, Aurangabad, pp 338, (1984).

Shashikant and V.K. Anand, Inter relationships of phytoplankton and physical factors in Mansar lake *Indian J.Ecol.* 5(2): 134-40, 1978b.

Tucker, C.S. amd Loyd, S.W. Phytoplankton communities in channel cat fish ponds. *Hydrobiologia.* 112: 137-41, 1984.

Vyas, L.N. and H. D. Kumar, Studies on the phytoplankton and other algae of Indrasagar tank, Udaipur, *Hydrobiologia* 31 (3/4): 42-34, 1968.

7

Hydrobiological Studies of Godavari River from Nashik Region (Maharashtra)

Resham Bhalla

ABSTRACT

The water of Godavari River is used for various activities like bathing and washing of clothes by large number of pilgrims. Other activities like washing of vehicles, throwing of vegetable peels and domestic sewage has deteriorated the water quality of holy Godavari river. The water samples were collected seasonwise (winter, summer, monsoon) during November 2002 to October 2003 to monitor physico-chemical and biological parameters like temperature, turbidity, total dissolved solids, pH, nitrates, nitrites, chlorides, sulphates, phosphates, dissolved oxygen, Phytoplanktons and Zooplanktons. Seasonal variations were noted and the results revealed that water of river is polluted.

Keywords: Hydrobiological studies, Seasonal changes, Godavari river.

Introduction

Water is life, so central to human life, yet over one billion people across the world have no access to safe drinking water. This is because water resources are under pressure and in danger as a result of potential pollution and contamination risks due to overuse and misuse of the resources in the world. Surface water is used for drinking purpose for human beings and other animals. Godavari, the biggest river in Maharashtra originates from Brahmagiri hills of Trimbakeshwar, at a distance of 29 km. from Nashik. River water is generally polluted by effluent discharges from nearby industrial plants as well as domestic wastes from local settlements situated on the bank of the river. Population and industrial growth and agricultural development have incurred a steadily increasing demand for water and this have resulted in a reduction in water quality and quantity. Due to large scale bathing activity, vehicle washing in the holy Godavari river, a significant rise in pollution of the river is observed. It is also learnt that the information regarding the hydrobiological studies of Godavari River is very scanty. Hence, Godavari river water was taken into consideration to examine its water quality in terms of physico- chemical and biological parameters, to evaluate the impact of various human activities in the catchment of the river.

Materials and Methods

In order to study the hydrobiological aspect of Godavari river three sampling stations were chosen as S1 (near Someshwar temple), S2 (Ramkunda) and S3 (near Tapovan). Water samples were collected from sampling station at an interval of one month representing seasonwise (winter, summer, monsoon) during Nov. 2002 to October 2003, in clean rinsed poly vinyl bottles to analyse the various physico-chemical and biological parameters according to the procedures given by APHA, AWWA and WPCF (1995). The phytoplanktons were studied using keys given by Adoni (1985).

Results and Discussions

The seasonal variations of physico-chemical and

biological parameters of three stations of Godavari River is presented in Table 7.1.

Temperature is an important parameters influencing biological reactions and chemical equilibrium in an aquatic ecosystem. Minimum temperature was recorded at stations S1 in winter (20.10°C) and maximum at stations S2 and S3 in summer (28.85°C) during the study period. An increase in summer and decline in winter of surface water temperature indicated that there was a parallel variation between air and surface water temperatures in the river. Similar observations were of Kannan and Jog (1980) in tropical impoundments.

Turbidity plays an important role in the energy dynamics of an aquatic ecosystem. The turbidity also varies with season. Maximum turbidity (163.25) was observed in summer at station S3 and minimum (95.25) in winter at station S1. During summer the turbidity was due to less flow of water, enriching organic matter. Sexsena et. al. (1966) reported highest value of turbidity in the river Ganga at Kanpur during rainy season due to surface runoff, high velocity of current and other impurities.

Total Dissolved Solids are generally of inferior palatability and may induce an unfavourable reaction in the transient consumer. The maximum concentration of TDS was in summer (274.50) at station S3 and minimum was found in winter (173.75) at station S1, probably because of stagnation and settlement of suspended particles. The high values of TDS may influence the taste, hardness and corrosive properties of the water (EPA, 1976).

The **pH** of Godavari River water remaining alkaline throughout the study period. Maximum pH was recorded in summer (7.80) at station S3 and minimum in winter (7.49) at station S1 with slight increase in monsoon at three stations. Srivastava et. al. (2003) reported that the value of pH ranged from 8.0 to 9.0 in Indian rivers. Ragothaman and Jaiswal (1995) observed pH of Tapi river in the range of 6.50 to 8.30.

Water bodies polluted by organic matter exhibit higher values of **Nitrates** (Kodarkar, 1995). Higher concentration of nitrate is an indication of organic pollution and eutrophication (Krishnan et.al., 1999). Concentration of nitrates were maximum in winter (2.87 mg/lit.) at station S3 and minimum in summer (0.63 mg/lit) at station S1. Leaching of fertilizers in winter used for agricultural purposes in the area covered for study is responsible for higher concentration in winter.

Higher level of **Nitrite** is a clear indication of large amount of faecal matter coming from domestic waste. Since nitrites are readily oxidized to nitrates are seldom present in significant concentration in surface waters. Maximum concentration of nitrites was observed in summer (0.06 mg/lit) at station S3 and minimum in monsoon (0.03 mg/lit at station S1. Jha and Barat (2003) observed higher values of nitrites, which may be due to oxidation of ammonia.

Maximum **Chloride** content was observed in summer (26.15) at station S3 and minimum in winter (13.23) at station S1. The rise of chloride in summer was due to rise in temperature and evaporation. Results are in agreement with Goel et. al. (1980). Ajmal et. al. (1985) found maximum value of chlorides during winter in Kalinadi. Venkateswarlu and Jayanti (1968) concluded that the values of chloride are associated with the organic pollution in some rivers and reservoirs, the main source of which is sewage.

Sulphates indicate the pollution of water by domestic wastes. Maximum concentration of sulphates were recorded in summer (51.48) at station S1 may be due to high temperature, high rate of evaporation and minimum in winter (28.80) at station S1 may be due to reduction of sulphate into sulphite and their consequent escape in the form of H_2S gas. The recommended upper limit for sulphate ions in water intended for human consumption is 250 mg/lit (ISI,1991).

Maximum **Phosphates** were observed in summer (1.52)

at station S3 and minimum in winter (0.83) at station S1, in Godavari River water. Phosphates are usually present in low concentration in natural unpolluted rivers. Many Esther and Johnson (1991) observed seasonal variations in two lakes of Hyderabad.

Maximum concentration of **Total Hardness** observed in winter (146.75) at station S3, followed in monsoon (117.50) at station S3 and minimum in summer (74.50) at station S1. The increase in hardness of water during winter and rainy season can be ascribed to greater amount of leaching of rocks and dissolution by flood waters. Hardness of river Ganga at Patna ranged from 80 to 180 mg./lit. (Ghose and Sharma, 1988).

Maximum **Dissolved Oxygen** was recorded in winter (7.48) at station S1 and minimum in summer (3.90) at station S3 during the study period. Similar observations were reported by Adebisi 1981 and Mitra 1982. The low oxygen content during summer may be due to low water. High temperature and decay of macro vegetation.

Phytoplanktons were reported maximum (2365.25) in summer at station S3 and minimum (326.50) in winter at station S1 may be due to cumulation effect of many factors appeared to govern their growth and occurrence. Jayaraju et. al. (1994) reported variations in Phytoplanktons and Zooplanktons with reference to physico-chemical parameters.

Zooplanktons were reported maximum in summer (908.25) at station S3 and minimum (203.75) in winter at station S1 during the present investigation. The difference may be due to seasonal variations the similar results observed by Saksena and Sharma (1987) and Haque et. al. (1988).

On the basis of above hydrobiological parameters it is concluded that the discharge of treated and untreated sewage, agricultural runoff and the holy activities being carried out may be responsible for degradation of water quality of holy Godavari River.

Table 7.1 : Seasonal variations of physico-chemical and biological parameters of Godavari river, from Nashik reaion

Parameters	W	S	M	W	S	M	W	S	M
Temperature (°C)	20.10	28.65	25.20	20.37	28.85	25.32	20.40	28.85	25.40
Turbidity (NTU)	95.25	143.50	131.75	104.00	149.75	134.50	107.50	163.25	142.50
TDS (mg/lit)	173.75	261.50	219.75	184.00	268.75	225.00	189.25	274.50	232.25
pH	7.49	7.70	7.58	7.53	7.55	7.50	7.65	7.80	7.75
Nitrates mg./lit.	2.55	0.63	1.51	2.73	0.67	1.59	2.87	0.69	1.65
Nitrites mg./lit.	0.04	0.03	0.03	0.05	0.05	0.04	0.06	0.05	0.05
Chloride (mg/lit)	13.23	25.00	17.33	13.78	25.75	18.06	13.53	26.15	18.35
Sulphates (mg/lit)	28.80	50.58	36.88	29.09	51.00	37.03	29.89	51.48	37.18
Phosphates (mg/lit)	0.83	1.45	0.98	0.88	1.48	1.03	0.89	1.52	1.03
Total Hardness (mg/lit)	125.00	74.50	99.50	139.25	83.50	108.75	146.75	88.25	117.50
Dissolved Oxygen (mg/lit)	7.48	4.40	4.93	7.20	4.13	4.65	6.93	3.90	4.58
Phytoplanktons(org/m)	326.50	2170.50	1230.25	370.00	2256.00	1320.00	424.50	2365.25	1375.75
Zooplanktons (org/ml)	203.75	884.00	370.00	214.75	880.75	375.50	220.50	908.25	377.25

Graphical representation of Seasonal variations of Physico-chemical and Biological parameters of Godavari river from Nashik region

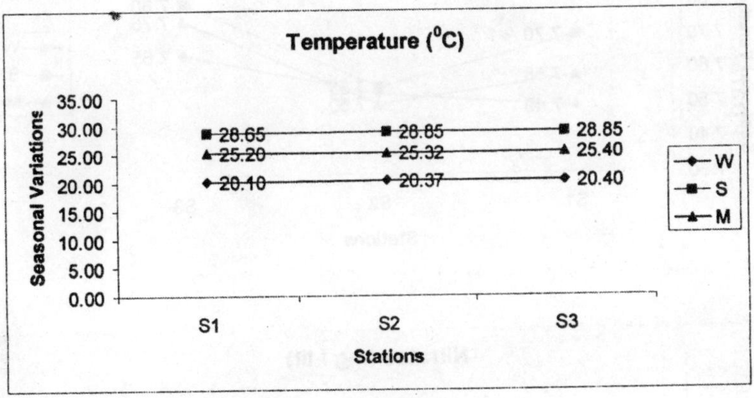

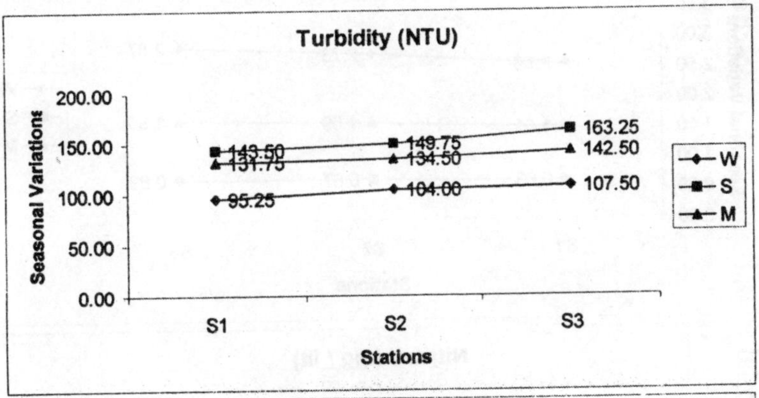

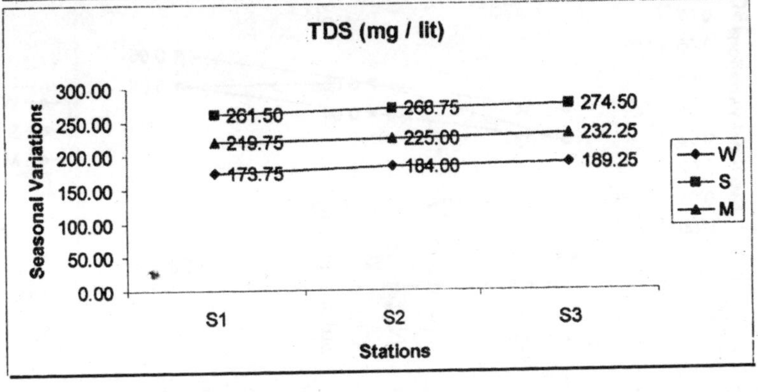

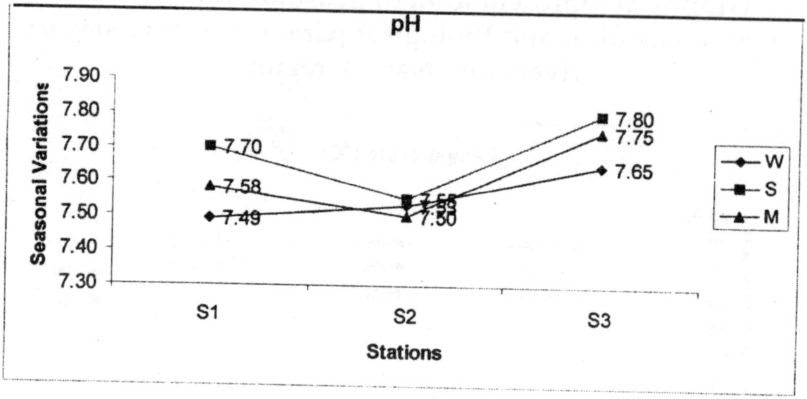
pH
Seasonal Variations
7.90
7.80
7.70
7.60
7.50
7.40
7.30
7.70
7.58
7.49
7.80
7.75
7.65
7.50
S1
S2
S3
Stations
W
S
M

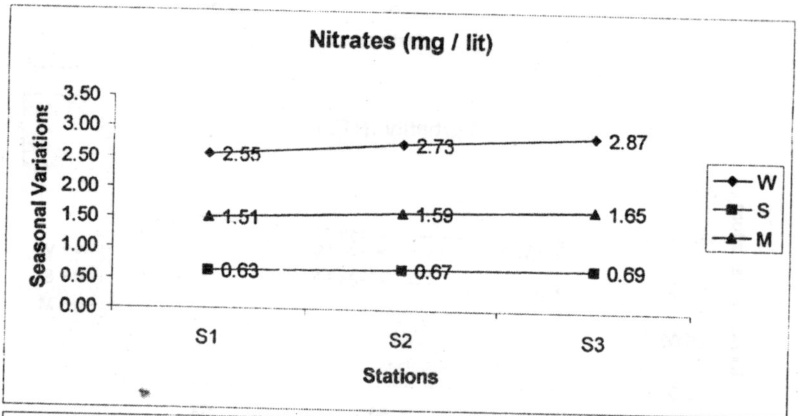
Nitrates (mg / lit)
Seasonal Variations
3.50
3.00
2.50
2.00
1.50
1.00
0.50
0.00
2.55
2.73
2.87
1.51
1.59
1.65
0.63
0.67
0.69
S1
S2
S3
Stations
W
S
M

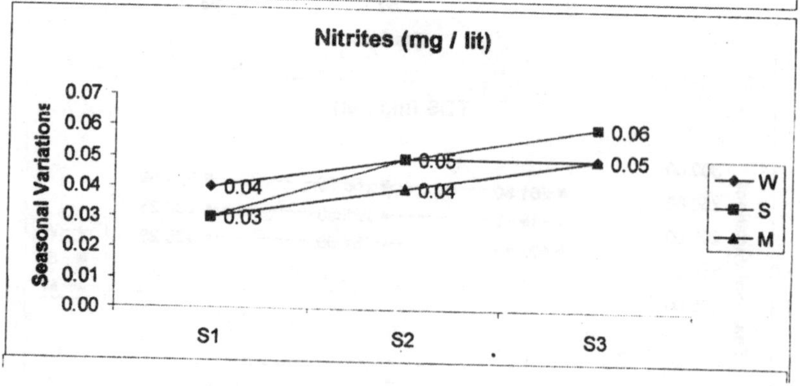
Nitrites (mg / lit)
Seasonal Variations
0.07
0.06
0.05
0.04
0.03
0.02
0.01
0.00
0.04
0.03
0.05
0.04
0.06
0.05
S1
S2
S3
W
S
M

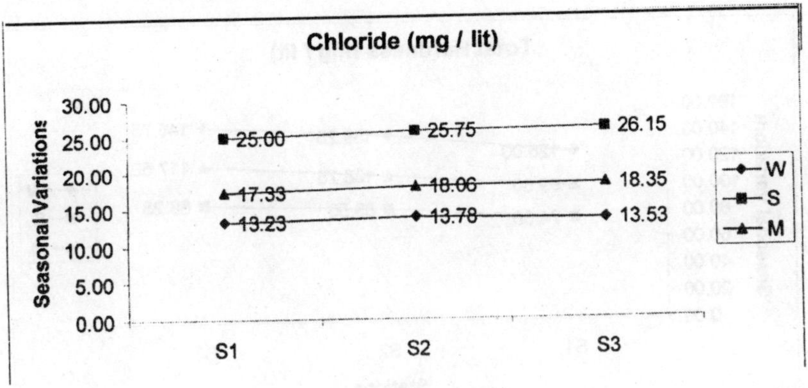
Chloride (mg / lit)
Seasonal Variations
30.00
25.00
20.00
15.00
10.00
5.00
0.00
25.00
25.75
26.15
17.33
18.06
18.35
13.23
13.78
13.53
W
S
M
S1
S2
S3

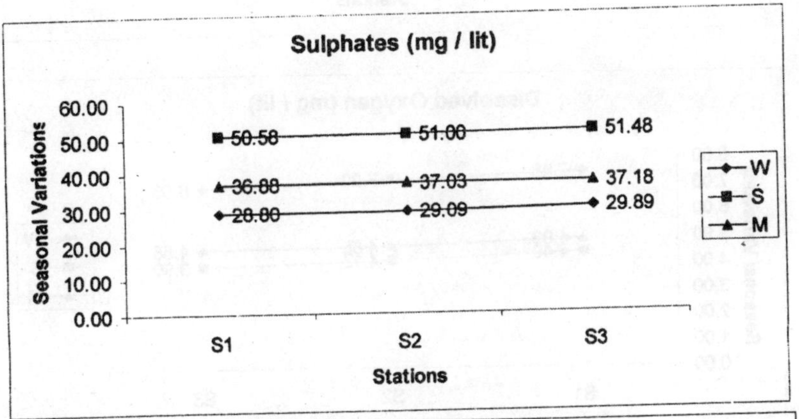
Sulphates (mg / lit)
Seasonal Variations
60.00
50.00
40.00
30.00
20.00
10.00
0.00
50.58
51.00
51.48
36.88
37.03
37.18
28.80
29.09
29.89
W
S
M
S1
S2
S3
Stations

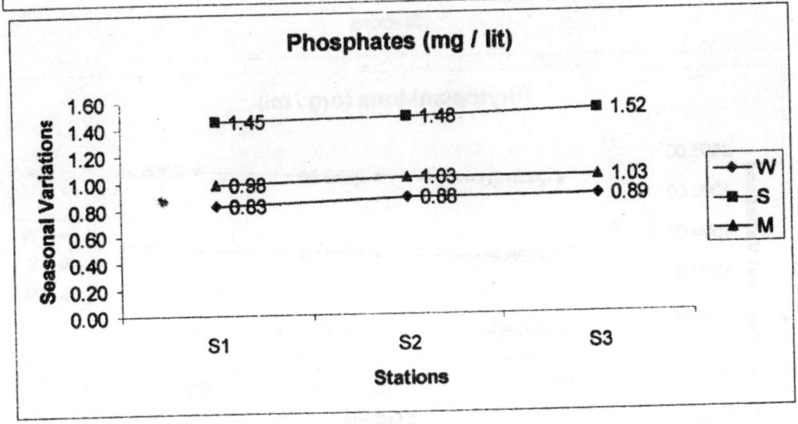
Phosphates (mg / lit)
Seasonal Variations
1.60
1.40
1.20
1.00
0.80
0.60
0.40
0.20
0.00
1.45
1.48
1.52
0.98
1.03
1.03
0.83
0.88
0.89
W
S
M
S1
S2
S3
Stations

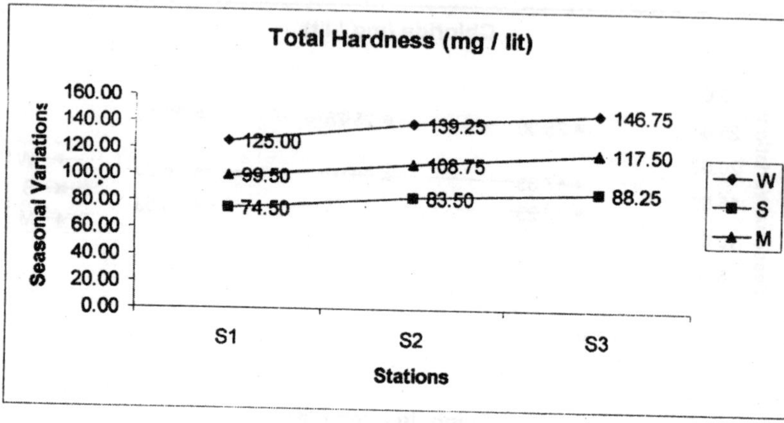
Total Hardness (mg / lit)
Seasonal Variations
160.00
140.00
120.00
100.00
80.00
60.00
40.00
20.00
0.00
125.00
139.25
146.75
99.50
108.75
117.50
74.50
83.50
88.25
W
S
M
S1
S2
S3
Stations

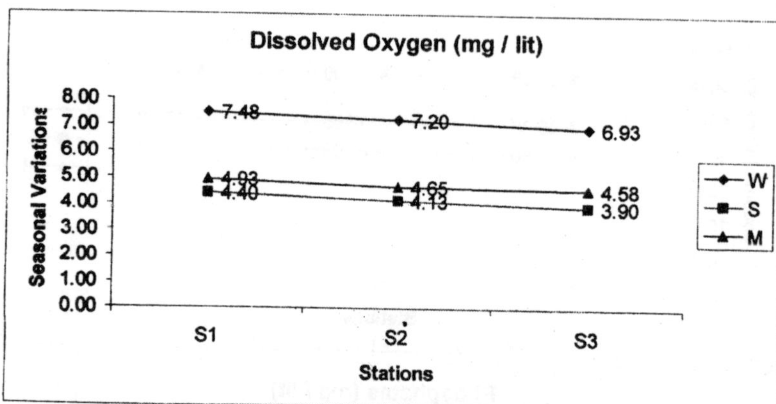
Dissolved Oxygen (mg / lit)
Seasonal Variations
8.00
7.00
6.00
5.00
4.00
3.00
2.00
1.00
0.00
7.48
7.20
6.93
4.93
4.65
4.58
4.40
4.13
3.90
W
S
M
S1
S2
S3
Stations

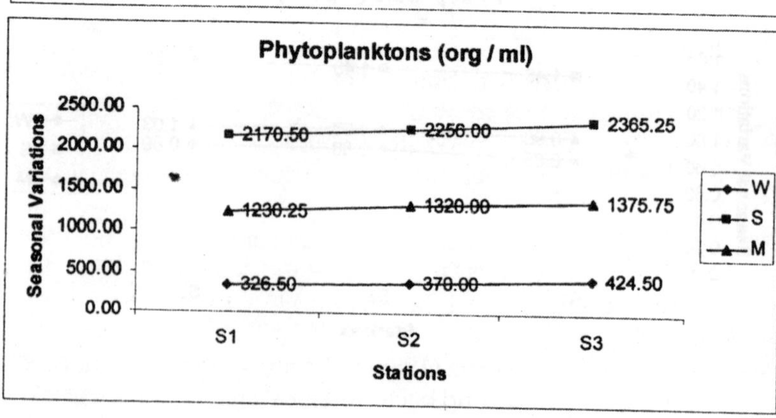
Phytoplanktons (org / ml)
Seasonal Variations
2500.00
2000.00
1500.00
1000.00
500.00
0.00
2170.50
2256.00
2365.25
1230.25
1320.00
1375.75
326.50
370.00
424.50
W
S
M
S1
S2
S3
Stations

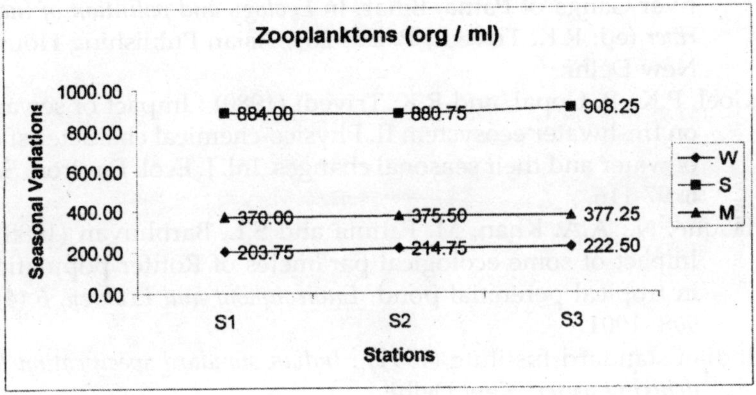

Acknowledgement

The author is thankful to co-ordinator, MG Vidyamandir and Principal LVH Mahavidyalaya for providing necessary research laboratory facilities.

References

Adebisi, A.A. (1981): The physico-chemical hydrology of a tropical seasonal river upper Ogun river Nigeria. *Hydrobiologia,* 79 (2): 157-65.

Adoni, A.D. (1985) : *Workbook on limnology,* Indian MAB committee, Dept. of Environment, Govt. of India.

Ajmal, M., Raziuddin and Khan, A.U. (1985) : Physico-chemical aspects of pollution in Kalinadi *JAWPC TECH., ANNUAL,* 12:106-14.

APHA, AWWA and WPCF : Standard methods for the estimation of water and waste water. American public Health Association, American Waste Water Association and Water Pollution Control Federation. 1 9th Edition Washington. D.C. (1995).

EPA (1976) : *Quality criteria for water,* Environmental protection Agency, Washington, D.C. Indian Standards (IS: 10500-1991 with amendment No.1 January 1993). Indian Standard Institute.

Ghose, N.C. and C.B. Sharma (1988) : The effect of drain water on the physico-chemical and bacteriological characteristics of the river Ganga of Patna, Bihar. In *Ecology and pollution of India river* (ed. R.K. Trivedi) P. 255-269, Asian Publishing House, New Delhi.

Goel, P.K., B. Gopal, and R.K. Trivedi (1980) : Impact of sewage on freshwater ecosystem II. Physico-chemical characteristics of water and their seasonal changes. Inl. J. Ecol. Environ. Sci. 6: 97-116.

Haque, N., A.A. Khan, M. Fatima and S.L. Barbhuyan (1988) : Impact of some ecological parametes of Rotifer population in tropical perennial pond. *Environment and Ecology,* 6 (4) : 998 -1001.

Indian standard Institute (1991) : *Indian standard specification for drinking water,* New Delhi.

Jayaraju, P.B., G.D.V. Prasadrao and S.V. Sharma (1994) : Seasonal variations in physico-chemical parameters and diversity in the flora and fauna of the river Munneru. A tributary to river Krishna, Andhra Pradesh, India *J. Aqua Biol.* vol. 9 (1 & 2): 198-220.

Jha, Prithwiraj and Barat, Sudip (2003) : Hydrobiological study of lake mirik in Darjeeling Himalayas J. Environ. Biol. 24 (3) : 339-344.

Kannan, V. and S.V. Jog, (1980): Diural depthwise and seasonal changes of physico-chemical factors in sathiyar reservoir, *Hydrobiologia,* 70: 103-17.

Kodarkar, M.S. (1995) *In conservation of lakes (with special reference to water bodies in and around Hyderabad*). IAAB Hyderabad, p. 82.

Krishnan, K., Thomas, Sabu., George, Sunil., Paul Murugan, R., Mundayoor, Satish and Das, M.R. (1999) : A study on the distribution and ecology of phytoplankton in the Kuttlanad wetland ecosystem, Kerala, *Poll. Res.* 18 (3) : 261-269.

Mary, Esther and Johnson Cynthia (1991) Nitrogen and phosphate contents of two freshwater lakes *J. Aqua Biol.* vol. 6 (1 & 2): 19-26.

Mitra, A.K. (1982) : Chemical characteristics of surface water at a selected guaging station in the river Godavari, Krishna and Tungabhadra. *Ind., J. Environ, Health,* 24 (2) : 165-79.

Ragothaman, G. and Jaiswal, R.N. (1994) : Studies on the hydrobiology of Tapi river from Jalgaon region (Maharashtra) with reference to phytoplankton. *Poll. Res.* 14 (2):181-94.

Saxsena, K.L., RN. Chakrabarty, A.K. Khan, S.N. Chattapadhyay, and H. Chandra (1966) : Pollution studies on the river Ganges near Kanpur, Environ. Health 8 (4) 270-285.

Saksena, D.N. and S.P. Sharma (1987) : Seasonal variations of copepode components of Zooplankton in a perennial tank, Janakal, Gwalior, India, *Acha Hydrochim. E. Hydrobiol.* 11(4): 479-84.

Venkateswarlu, T and TV. Jayanti (1968) : Hydrobiological studies of the river Sabarmati to evaluate water quality. *Hydrobiologia,* 33 (3-4) : 422-48.

8

Seasonal distribution of hydrobiological parameters in relation to primary productivity and their correlation of Godavari river water, Nashik

Resham Bhalla
Balwinder Sekhon

ABSTRACT

Seasonal distribution of various hydrobiological parameters such as temperature., pH, turbidity, conductivity, TDS, chlorides, nitrates, sulphates, phosphates, zooplanktons, phytoplanktos, DO, BOD, COD along with the primary productivity and their coefficient correlation has been studied at three stations A,B,C, during three seasons (Winter, Summer, Monsoon in the year November 2001 to October 2002). In the present study variations have been observed during the three seasons and physico-chemical parameters are showing direct correlation with the planktonic growth and primary productivity thus enriching the Godavari river with biomass.

Keywords : Seasonal distribution, Godavari river, hydrobiological parameters, coefficient correlation.

Introduction

Water pollution occurs due to presence of dissolved inorganic or organic materials which change the physico-chemical characteristics of water bodies and poses threat to aquatic life and organisms responsible for self-purification of water bodies. Planktons are important biotic components of aquatic habitat. The primary productivity of river has a great significance in improving the fertility of planktons. A definite interrelation exists between various factors such as light, turbidity, nutrient level, planktons in relation to productivity of the river. Biological mechanisms play a critical role in the purification of polluted water through treatment plants. The presence or absence of certain organisms can be used to determine the condition of water and efficiency of water treatment plants. Aquatic organisms constitute a vital link in the food chain in the aquatic ecosystem and its productivity directly depends on physico-chemical characteristics of water. Thus the study of limnological parameters assume greater importance for its practical application as it provides the most precise results and makes possible an insight into interaction among various physico-chemical factors.

> "Limnology is the study of functional relationship and productivity of fresh water biotic communities as they are affected by dynamics of physical, chemical and biotic environmental parameters." (Wetzal, 1975)

The Godavari River selected for the present investigation originates in the Nashik district at Trimbakeshwar in Maharashtra state. Godavari is the most celebrated river, not only of the district but also of the peninsular India. After running through a distance of 1465 km it joins the Bay of Bengal in Andhra Pradesh. It has a catchment area of 3,12,812 sq. km of which 69.3 per cent lies in Maharashtra.

Sampling sites

Station A : Gangapur (Near Someshwar Temple), the

point before the discharge of agro-industrial effluents. The water is used for agriculture, bathing and washing purpose and agricultural run off is also released.

Station B: Ramkund also known as Gangaghat is a holy place. Various rites and rituals are performed here. Water is used for bathing and washing also. The domestic waste water is largely released before and after this point.

Station C : Tapovan area near Laxminarayan temple, before this point sewage and some industrial waste is also released into water.

Materials and Methods

Water samples were collected in between 8 a.m. to 10 a.m. regularly for a period of one year at an interval of one month from November 2001 to October 2002 from three stations, i.e. Station 'A' Station 'B' and Station 'C'.

The sample were collected, labelled properly and brought to the laboratory for further analysis. Conductivity was measured by using conductivity meter.Turbidity was measured by using Nephalometer (Turbiditimeter). Total Dissolved solids were estimated by gravimetry. The methods used for chemical analysis were standardized according to the procedures given in "Standard Methods for the Examination of Water and Waste Water" by APHA-AWWA-WPCF (1989), the constituents are reported in milligrams per litre Flora and Fauna were identified using keys given by Sehgal (1983); Adoni, (1985); APHA (1985). Primary productivity was estimated by using the changes in dissolved oxygen concentration in dark and light bottle by following the method of Gaarder and Gran (1927). The data were subjected to statistical test for correlation, computation to understand the complex eco-interactions. To study the correlation and influence of physico-chemical variables on plankton density, simple co-efficient correlation of the variables taken by pairs and analysis of independent variables in relation to dependent and independent variables were carried out.

Results & Discussion

Temperature : Water temperature is recorded lower in monsoon may be due to cloudy weather and influx of rain water and winter due to extreme cold and shorter sunshine period whereas higher in summer followed a common pattern as observed by Swarnlatha, 1994; Rajasegar, 2003.

pH : Maximum pH was recorded in summer at Station 'C' and minimum in winter at Station 'A' with slight increase in monsoon season at Station 'C'. Similar trend has been reported by Sreenivasan (1965); Vijayaraghavan (1971) and Gupta and Mehrotra (1986).

Turbidity: Turbidity values were highest in summer season due to natural runoff and waste followed in monsoon and least in winter. Mathew et. al. (1972) also observed order of turbidity as summer > rainy > winter in a rural pond near Mandala (M.P.).

Conductivity: The conductivity values low during winter and high during summer at three stations. A significant positive correlation was observed between conductivity and temperature at three sampling stations. An average seasonal conductivity of river water varied due to ionic composition of water, rainfall, biota, etc., and variations in ground discharges also influence the conductivity. This is also opined by Datta et. al. (1988).

Total Dissolved Solids (TDS): The maximum concentration of TDS was in summer while the minimum value was found in winter, probably because of stagnation and settlement of suspended particles. In the present investigation the high values of TDS were mainly due to different types of industrial wastes, activities of slum dwellers and addition of municipal sewage to the Godavari river water at Nashik.

Chlorides: The rise of chloride in summer was due to rise in temperature and evaporation thereby affecting the process of photosynthesis. In monsoon, the lower values are due to rainfall leading to dilution while the rise in winter is correlated with phytoplanktons. Results are in agreement with Goel et. al. (1980).

Nitrates: Concentration of nitrates were highest in winter, followed in summer and least in rainy season. The observed maximum values of nitrates are in agreement with Prasad and Saxena, (1980); Rana, et. al. (1991).

Sulphates: Values of sulphates ranged minimum in winter at Station 'A' and maximum in summer at Station 'C'. The maximum amount recorded in summer may be due to high temperature, high rate of evaporation and reduction in water level.

Phosphates: The maximum phosphates are in summer and minimum in winter in the year 2001-02. The quantities of phosphates varied throughout the investigation at all stations of the Godavari river. Similar findings were observed by Welch (1952) and Hutchinson (1957).

Zooplanktons: The seasonal variations show direct relationship with physico-chemical parameters such as pH, temperature, DO, CO_2, phosphates etc., indicating that there is direct relationship between phytoplanktons and zooplanktons. In winter, small growth of phyto and zooplanktons may be due to low temperature.

Phytoplanktons: The pH, temperature, intensity of light penetration, organic matter may be responsible for the phytoplankton production.

Dissolved Oxygen (DO): The maximum Dissolved Oxygen (DO) was recorded in winter at Station 'A' and minimum Dissolved Oxygen (DO) was recorded in summer at Station 'C'. Vyas (1968); Deshmukh (1964) showed an inverse relationship of Free CO_2 with DO. Bahura (1998) reported an inverse relationship of DO with temperature. The low oxygen content during summer may be due to low water level, high temperature and decay of macro-vegetation.

BIOCHEMICAL OXYGEN DEMAND (BOD): The BOD values were found high in summer followed by monsoon and low in the winter, similar observations were made by Qumerunnisa (1985).

Chemical Oxygen Demand (COD): Higher values of COD indicate the higher microbial activities and presence of oxidizable organic matter. Khulbe and Durgapal (1993) have reported similar findings.

Primary Productivity: The results of productivity study of Godavari river at Nashik are as such

Maximum	*Minimum*
GPP 283.27(mg *C/m²Ihr)*	113.16(mg *C/m²/hr)*
NPP 211 .53(mg *C/m²/hr)*	86.28 (mg *C/m²/hr)*
CR 61.89(mgO2/1/hr)	26.88(mgO2/1/hr)

Varghese (1992) also reported maximum productivity during summer followed by winter and rainy season. Seasonal variations and changes in physico-chemical and biological parameters also had an influence on productivity.

Co-Efficient Correlation: The water temperature was positively correlated with conductivity. These results are in agreement with Singhai et. al. (1990).The temperature is known to play a prime role in the productivity of water. However in the present work maximum number of planktons were recorded during summer season. Both phyto and zooplanktons showed a positive correlation with temperature. Temperature showed positive correlation with BOD and COD whereas BOD and COD are directly proportional but show inverse relation with Dissolved Oxygen.

Summary and Conclusion

The overall study of physico-chemical, microbial and biological parameters of Godavari river near Nashik reveals that the water quality is changing gradually due to certain anthropogenic activities in and in the vicinity of river. Further, the discharge of treated and untreated sewage, agricultural run off may also be responsible for changing the water quality, which reduces the utility for domestic use.

Table-8.1 : Hydrological Parameters in relation to Primary Productivity and their Correlation

Parameters	Station A			Station B			Station B		
	Winter	Summer	Monsoon	Winter	Summer	Monsoon	Winter	Summer	Monsoon
Temperature (0)	20.00	28.50	25.02	20.12	28.77	25.20	20.12	28.72	25.15
pH	7.38	7.68	7.78	7.45	7.70	7.80	7.55	7.80	7.90
Turbidity (NTU)	94.75	139.50	131.00	98.75	148.00	137.00	103.50	158.25	147.00
Conductivity (umho/cm.)TDS (mg/lit)	218.83	304.96	289.68	229.38	318.73	306.23	239.93	331.60	320.20
	174.75	253.75	206.00	183.25	265.00	216.25	186.75	269.06	224.75
Chlorides mg/lit	13.25	24.58	17.47	13.40	25.11	18.15	13.63	25.40	18.34
Nitraties (mg/lit)	2.58	0.63	1.48	2.73	0.65	1.55	2.80	0.67	1.60
Sulphates (mg/lit)	28.56	48.55	36.58	28.74	49.68	36.90	28.80	49.28	36.95
Phosphates (mg/lit)	0.83	1.42	1.01	0.85	1.44	1.03	0.85	1.48	1.05
Zoo-Planktons (mg/ml)	199.25	857.50	376.00	206.50	879.25	374.50	217.00	893.00	375.50
Phyto-Planktons (mg/ml)	318.50	2011.00	1199.00	331.50	2115.00	1282.75	354.50	2219.75	1354.00
DO (mg/lit)	7.48	4.23	5.00	7.05	4.20	4.68	6.90	3.88	4.53
BOD(mg/lit)	3.98	9.58	10.20	4.05	9.90	10.60	4.03	10.13	11.00
COD m flit	49.05	62.93	73.23	51.25	64.93	75.88	52.10	65.80	75.73
GPP (mg/m^3 /hr)	119.61	266.78	113.16	130.93	277.49	118.57	137.02	283.27	121.51
NPP (mg/m^3/hr)	87.76	211.53	86.28	97.61	219.07	88.62	101.03	221.27	89.70
CR(mgC/m^3 /hr)	31.83	55.25	26.88	33.31	59.41	29.94	35.99	61.89	31.81

Table 8.2 : Co-Efficient Correlation

Parameters	Station A	Station B	Station C
Nitrates and dissolved oxygen (DO)	0.7928	0.7959	0.8173
Turbidity and total dissolved Solids	0.7009	0.8067	0.6721
Temperature and Conductivity	0.8681	0.8730	0.8680
Temperature and pH	0.6635	0.5627	0.6189
Temperature and chlorides	0.9384	0.9544	0.9509
Temperature and BOD	0.8518	0.8491	0.85 19
Temperature and COD	0.5520	0.5324	0.5757
Temperature and Phytoplanktons	0.8967	0.8954	0.8959
Temperature and Zooplanktons	0.8426	0.8561	0.8477
Sulphates and phytoplanktons	0.8612	0.8732	0.8635
Sulphates and zooplanktons	0.8833	0.8993	0.89 39
Phosphates and phytoplanktons	0.8740	0.8746	0.8699
Phosphates and zooplanktons	0.9341	0.9248	0.9278
Temperature and GPP	0.7567	0.7290	0.7180
Temperature and NPP	0.8008	0.7315	0.7571
Temperature and CR	0.4001	0.3760	0.1360

References

Adoni, A.D. (1985) *Workbook on limnology,* Indian MAB Committee, Deptt. of Environment, Govt. of India.

APHA, American Public Health Association (1985) Standard method for the examination of water and waste water. 14th edu. APHA-AWWA-WPCF. Washington, D.C. p. 1193.

American Public Health Association (APHA), American Water Works Association (AWWA) and Water Pollution Control Federation (1989) : Standard Methods for the examination of Water and Waste Water, (ed. L.S. Clesceri, A.E. Greeenberg and R.R. Trussell (l7th edu.), APHA, Washington D.C.

Baruah, A.K., R.N. Sharma and G.C. Borah (1993) Impact of sugar mill and distillery effluents on water quality of river Gelabil, Assam. *Indian J. Environ, Hlth, vol. 35 (4)* :288-293.

Datta, N.C., B.K. Bandyapadhyay, A. Majumber and D. Ahuja (1988) Hydrological profile of Hooghly sector (Bally to Bandel) of river Ganga. In Ecology and *Pollution of Indian river.* R.K. Trivedy (Ed). 5: 113-129.

Deshmukh, S.B., N.S. Phodke and Kothandaraman (1964) : Physico-chemical characteristics of Ambazari lake water. *Ind. Joun. Hlth.* 6 (3) 186-88.

Gaarder, T. and H.H. Gran (1927) : Investigation of productivity of plankton in the Osio ford. *Rapop Cons. Explor. Mar.* 144: 56-160.

Goel, P.K., B. Gopal and R.K. Trivedi (1980) : Impact of sewage on freshwater ecosystem-II physico-chemical characteristic of water and their seasonal changes. *In!. J Ecol. Environ. Sl.* 6: 97-116.

Gupta, A.K. and R.S. Mehrotra (1986): Studies on seasonal variation in pH and dissolved oxygen content in Sanhit Sarovar Kurukshetra. *Geobios.* 13 (6) : 276- 78.

Hutchinson, G.E. (1957) : *A treatise on limnology,* vol. 1. Geography, physics and chemistry. New York. John Wiley and Sons. Inc 1015.

Khulbe, R.D. and A. Durgapal (1993) : Evaluation of drinking water quality at Bhimtal, Nainital, Uttar Pradesh. *Poll, Res. 21(2):109-111.*

Mathew, V., A. Chauvan and L.P. Naik (1972) : Hydrobiological studies of a domestically polluted tropical pond I, physico-chemical characteristics. *Pollution Research,* 11: 95-100.

Prasad, B.M. and Manjula Saxena (1980) : Ecological study of blue-green algae in river Gomti, India. *J. Environ. Hlth,* 22(2) : 151-68.

Quinerunnisa (1985) : Ecobiology of freshwater ciliates, Ph.D. Thesis, Marathwada University, Aurangabad, MS., India.

Rajasegar, M. (2003) : Physico-chemical characteristics of the Vellar estuary in relation to shrimp farming. *J. Environ. Biol.* 24 (1) : 95-101.

Rana, B.C., R.P. Vijayvegia, and Y.D. Tiaglli (1991) : Limnology of a eutrophic lake Udaisagar, Udaipur, India. *Indian J. Ecol.,* 18(20): 161-65.

Sehgal, K.L. (1983): Planktonic copepods of fresh water ecosystem Interprint, New Delhi.

Singhai, S., G.M.A. Ramani and U.S. Gupta (1990) : Seasonal variations and relationship of different physico-chemical characteristics in the newly made Tawa. *Limnologia (Berlin)* 21(1): 293-301.

Sreenivasan, A. (1965) : Limnology and productivity of tropical impoundment, Amravathi reservoir, Madras State, India. *Hydrobiologia,26: 501-16.*

Swarniatha, N. (1994) : Seasonal rhythm of various physico-chemical characteristics of Saroornagar lake. *J. Mendel.* 11 (1&2) 65-66.

Varghese, Mathew (1992) : Seasonal primary production in the Macrophytic community of Govindgarh lake, Rewa (MP). *J. Ecobiol., 4* (1):11-14.

Vijayaraghvan, 5. (1971) : Seasonal variation in primary productivity in three tropical stations. *Hydrobiologia, 38 : 1 (3-4): 395-408.*

Vyas, L.N. and H.D. Kumar (1968) : Studies on the phytoplankton and other algae of Indira sagar tank, Udaipur, India. *Hydrobiologia,*31 : 421-34.

Welch, Paul S. (1952) : *Limnology,* 2nd edu. McGraw-Hill Book Co., New York, 1-538.

Wetzel, R.G. (1975) : Limnology, W.E Saunders Co Philadelphia, London, pp. 743.

9

Study of Physico-Chemical properties of the Pomegranate Soil of Malegaon Tehsil Nashik (Maharashtra State)

B.S. Yadav
A.R. Hiray
Imran Quraishi

ABSTRACT

The physico-chemical parameters of pomegranate soil of Malegaon tehsil was analysed. The results reveals that the soil is alkaline and has highest bulk and particle density. The water holding capacity was also high. The content of sodium and potassium was fluctuating and moderate. This type of soil is suitable for normal and healthy growth of the plants. The high porosity is favourable for the microbial growth and root propogation of the plants. In the present experiment moderate porosity of soil was observed. The result of the present investigation are shown in Table 9.1-9.2, and Graphs 9.1–9.8 reveals that there is a lots of variations among the soil samples collected from different pomegranate fields of Malegaon tehshil.

Introduction

Soil is the naturally occurring, unconsolidated or loose

covering of broken rock particles and decaying organic matter on the surface of the earth capable of supporting life. In simple farms soil has three components solid, liquid and gas. The solid phase is a mixture of mineral and organic matter. Soil particles pack loosely, forming a soil structure filled with voids. The solid phase occupies about half of the soil volume. The remaining void space contains water and air.

All soils initially come from rocks, this is termed the "parent material". In addition to the soil parent material, soil formation is also dependent upon other prevailing processes affecting soil formation. The soil formation process is termed "Pedogenesis" climatic conditions are important factors affecting both the form and rate of physical and chemical weathering of the parent material.

Soil is the main rooting medium for agricultural plants and long-term cultivation affects the fertility of soil and makes the cues of low yield of crops, therefor time to time monitoring or analysis of soil fertility parameters must be carried out to get the satisfactory yeild of pomegranate, which is cultivated on large scale in Malegaon tehsil of Nashik district, Maharashtra State India.

Material and Methods

For the present experiment, soil samples were collected from eight pomegranate fields of various villages of Malegaon tehsil. The sample collection sites are designated as PGF1, to PGF8 (Annexure 1). The soil samples were collected from 15 to 20 cm depth and thoroughly mixed on a clean piece of cloth and kept for drying on flat tray. Drying causes changes in the physical and chemical charecteristics of soil. When soil was air dried stones pieces and macro organic matters were picked out and weighed, after weighing samples were crushed and seived. Larger lumps were broken up by hand and then the soil is grinded by rolling gently with a wooden roller. After grinding, soil was screened through a 2 mm sieve then samples were packed in polythene bags which were tagged and lebeled properly and then used for analysis. Following parameters were analysed by using the appropriate methods described by APHA, Jalotra et. al., Hesse et. al., Tardon brown and Guggenheim et. al. (1995).

(a) Bulk density (b) Particle density
(c) Soil water content (d) Porosity
(e) Air filled porosity (f) pH
(g) Lime potential (h) Electrical conductivity
(i) Calcium carbonate (j) Organic matters
(k) Elemental determination (Sodium & Potassium)

Results

The results of the present investigation are presented in Table 8.1-8.2 and Graph 8.1-8.8. The present results reveals that the bulk density was highest in PGF1 (1.1882) followed by PGF6 (1.1513) and PGF2 (1.1242). The lowest bulk density was recorded in PGF5 (1.0007), this may be due to high water content of soil (Table 8.1 and Graph 8.1).

The particle density of PGF1 showed highest value (2.8902) followed by PGF2 (2.3474) and PGF4 (1.9940). The lowest particle density was recorded in PGF3 1.7525 (Table 8.1 Graph 8.1).

The high water holding capacity is good characteristic of soil. The highest water content was measured in PGF3 (19.048 per cent) followed by PGF5 (18.988 per cent) and PGF4(15.816 per cent). The lowest water content was recorded in PGF1 6.147% (Table-8.1 Graph 8.1)

Air filled porosity of PGF1 shows higher value (0.4511) followed by PGF2 (0.3494) than the other fields. The lowest value shown by PGF3 0.01268 (Table 8.1 Graph 8.3).

All the soil sample showed pH between the range of 8.84 to 9.24 which indicate that soil sample are alkaline in nature (Table 8.1 Graph 8.4).

The lime potential is used to determine the

carbondioxide concentration in soil air of the calcarious soil (Ulrich 1961).

The highest lime potential was measured in PGF7 (7.71) followed by PGF6 (7.7) and PGF4 (7.66). The lowest LP was measured in PGF1 which is 7.29 (Table 8.1 Graph 8.4).

The presence of cation and anion in the form presence of salt in the field influenced the EC of that field. All the soil samples were shown normal salt content which is suitable for normal and healthy growth of plants.

The EC value was recorded for PGF_6 (0.616) and followed by PGF7 (0.551) is highest then others; lowest value observed in PGF4 (0.162) (Table 8.1 Graph 8.5).

The highest value of organic carbon and organic matter was recorded in PGF3 (1.08 and 1.862) followed by PGF5 (1.07 and 1.845) and lowest in PGF1 (0.95 and 1.638) (Table 8.2 and Graph 8.7). The entire soil samples shows more than enough organic matter and hence there is no need to supply externally biofertilizers.

Percentage of $CaCO_3$ implies the calcarious and non-calcarious nature of soil samples shows per cent calcium carbonate in between 5.595 to 5.6 which implies that the samples are calcareous in nature (Table 8.1 and Graph 8.6).

Discussion

The important physical parameters in the soil analysis are pH and electrical conductivity. The analysis of samples shows the pH value ranging in between 8.84 to 9.33 which shows alkaline nature of soil. High alkalinity of the soil may be due to maximum use of fertilizer, but since several years farmers are using biofertilizers which may decrease the pH near to neutral for healthy growth of the plant Aggelides and Laoda (2000) reported the fluctuation in porosity, bulk density and pH of the soil Similar observation were made by Masle and Passioura (1987).

The chemical parameter which are actually govern the fertility of soil are OM, OC, N, P.K. In the present experiment, the content of K was sufficient in all soil samples and OC and OM was also in enough quantity. Prasana (1994) also noticed the similar results. EC is ranging from 0.616 to 0.162 in the present experiment which is normal and does not affect the growth of the plant. Katarki (2000) also noticed the reduced EC value with the application of amendments like FYM (Form Yord Manure) and green manure Kamalkumari and Singaram (1986) reported the benefits of FYM on stabilized soil structure. Bajwa and Josan (1989) reported the increased alkalinity with high pH of surface soil. Shrikanth et. al. (2000) reported that the alkalinity of the soil decrease by adding the enriched form waste compost enriched with weeds and cow dung slurry. Hedsa et. al. (2004) observed fluctuating organic carban and organic matter in all the soil samples. Further they reported that composted municipal waste increase the soil organic matter.

On the basis of the performance given by the collected soil samples the sample PGF1, PGF2, PGF6, PGF7 and PGF8 shows good characteristics for healthy propagation of pomegranate plants.

Acknowledgments

Authors are very much thankful to the Principal, M.S.G. Arts, Science and Commerce College, Loknete Vyankatraoji Hiray Marg, Malegaon-Camp-423 105 Dist. Nashik for providing the laboratory facilities.

References

.Bajwa, M.S. and A.S. Josan (1989) Effect of alternating sodic and non sodic irrigation on build up of sodium in the soil and crop yeild in north India Exp. Res. 25, 199-205.

Hedas, A., M. Agassi, H. Zhevile L. Kaustty, G.J. Levy, E. Fizik and M. Gotessman. (2004) Mulching with composted municipal solid waste in the central Negare, Israel II. *Soil tillage Res* 78, 115-28.

Hesse, P.R. Chapter 1 - Origin and nature of soil. Chapter II Preparation of soil sample for analysis. Chapter III, Soil reaction, pH value and lime potential. A textbook of soil chemical analysis.

Jalotra, S.K., Romes Khera and B. S. Ghyman, chapter 2, Properties of soil, methods in soil physics, Narsoa Publishing House, 16-61.

Kamalakumari, K. and P. Singgaram. (1996) Effect of continuous application of FYM and NPK as fertility status of soil, yeild and nutrient uptake in maize Madras Agril J. 83, 181-84.

Katarki, N (2000) Effect of organics and inorganics on growth and yeild of maize under irrigated condition. MSc (Agri.) thesis, University of Agriculture Sciences, Dharwad.

Hebsur, N.S., T.D. Bharamgoudar and H.M. Pradeep (2009) Effect of conjuctive use of canal and borewell water irrigation on maise yeild and soil properties at Ghataprabha commond area of Karnataka. *J.Ecotoxicol. Environment.* 19(1) 41-48.

Prasanna, S.M. (1994) Effect of direct and conjunctive use of saline water on wheat in TBP commond area, MSc thesis, University of Agriculture Sciences, Dharwad.

Shrikant, K (1997) Enrichment of compost and its effect on soil properties and crop growth. Ph.D. thesis, UAS, Banglore, India.

Tardon, H.L.S. Analysis of soil for available major nutrients, methods of analysis of soil plants, water and fertilizer.

Guggenheim, Stepen and R.T. Martin. (1995) "Definition of clay and clay mineral. Journal report of the AIPEA nomenclature and cms nomenclature committees" Clays and Clay Minerals 43, 255-56.

Table 9.1 : Physical properties of different PGF soil samples

Sample	Db	Dp	SWC%	F	Fa	pH	L.P.	EC	%$CaCO_3$
PGF1	1.1882	2.8902	6.1473	0.59	0.451	9	7.29	0.202	5.5835
PGF2	1.1241	2.3474	8.2792	0.52	0.349	8.84	7.36	0.2	5.6117
PGF3	1.1086	1.7525	19.048	0.37	0.013	9.33	7.5	0.193	5.5867
PGF4	1.0412	1.994	15.8169	0.48	0.194	9.24	7.66	0.162	5.59
PGF5	1.00079	1.7643	18.988	0.43	0.114	9.07	7.63	0.196	5.6034
PGF6	1.15133	1.7565	7.0739	0.34	0.193	8.95	7.7	0.616	5.595
PGF7	1.013	1.8936	14.7532	0.47	0.205	9	7.71	0.551	5.6
PGF8	1.1229	1.8202	12.4157	0.38	0.135	8.96	7.61	0.255	5.5884

Graph 9.1 : Comparison of Dp and Db

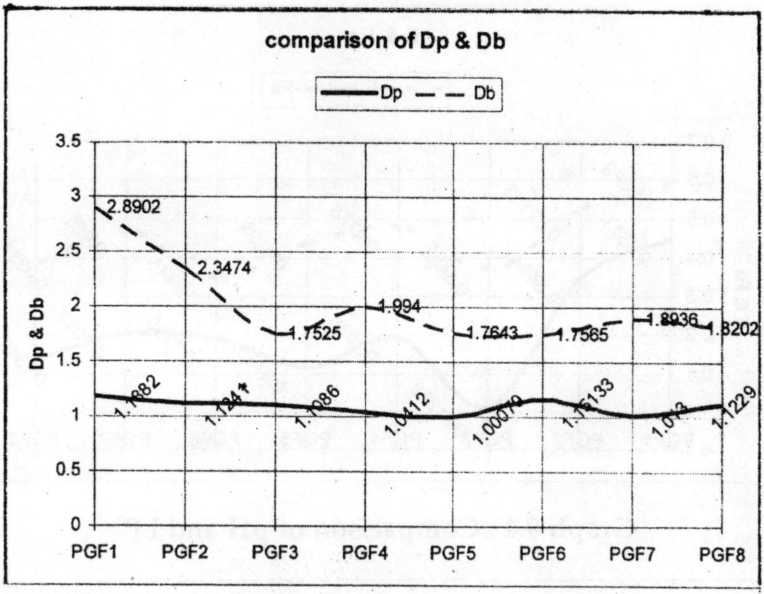

Graph 9.2 : Comparison of SWC

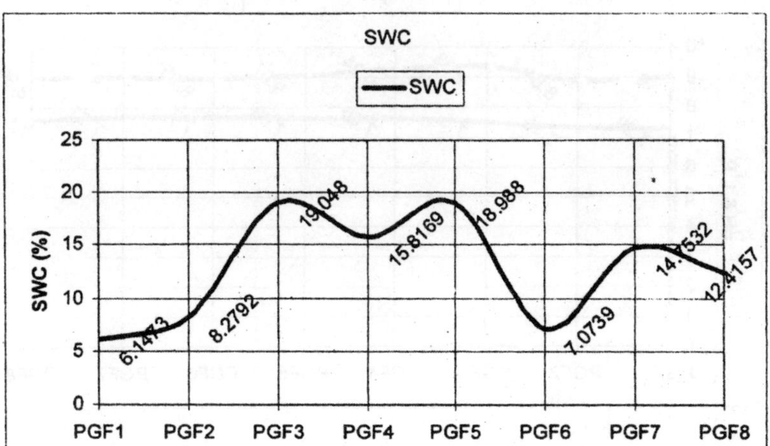

Graph 9.3 : Comparison of F and Fa

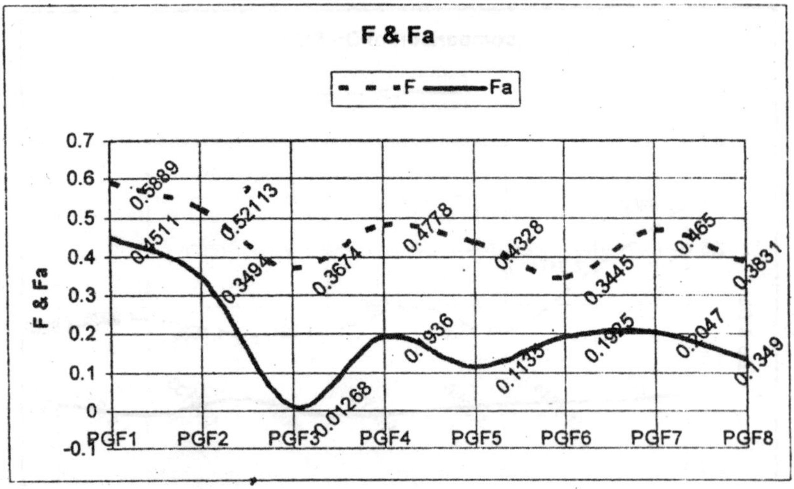

Graph 9.4 : Comparison of pH and LP

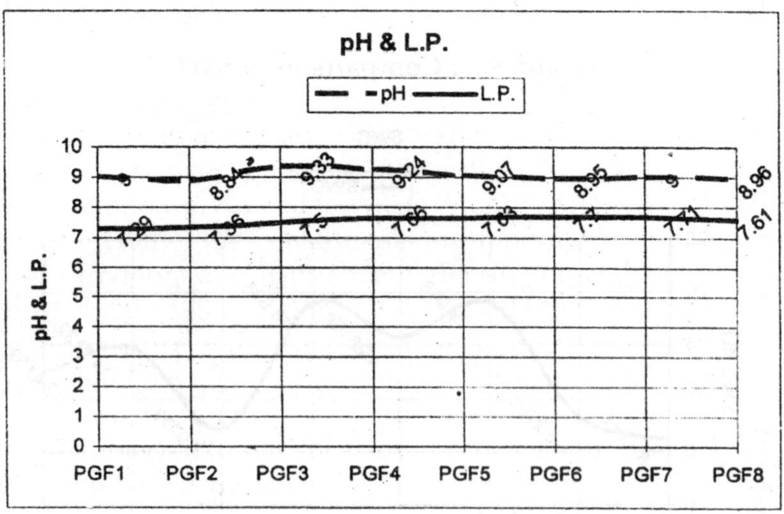

Graph 9.5 : Comparison of EC

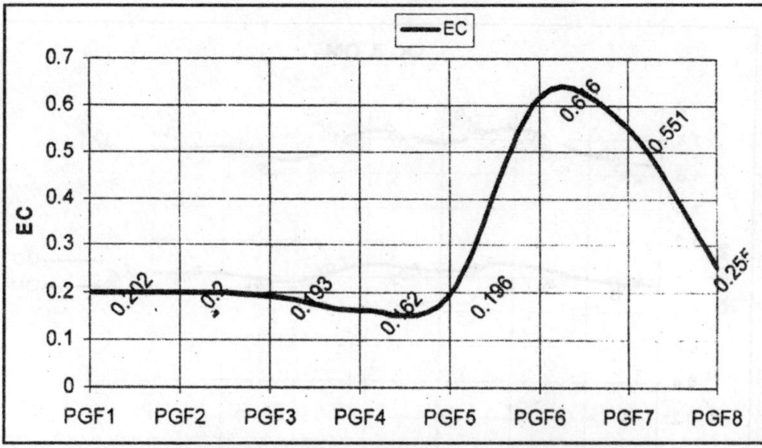

Graph 9.6 : Comparison of % $CaCO_3$

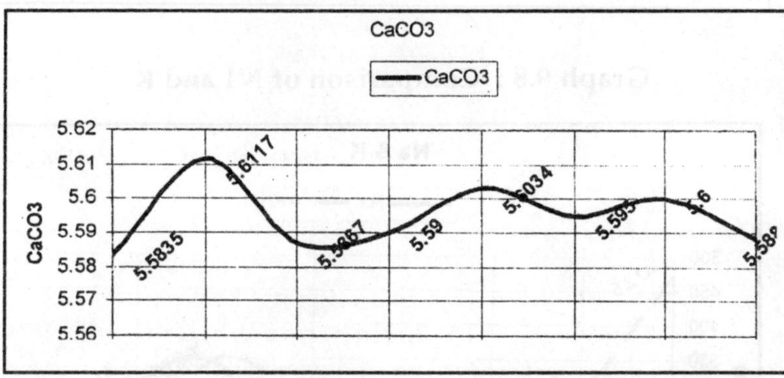

Table-9.2 : Chemical properties of different PGF Soil samples

Sample	OC	GM	Na	K
PGF1	0.95	1.6378	145.35	472.5
PGF2	0.99	1.7067	100.91	303.75
PGF3	1.08	1.8619	83.36	202.5
PGF4	1.03	1.7757	77.85	236.25
PGF5	1.07	1.8446	95.6	281.25
PGF6	0.97	1.6722	335.02	348.75
PGF7	1.01	1.7412	307.12	247.5
PGF8	1.01	1.7242	134.66	236.25

Graph 9.7 : Comparison of OC and OM

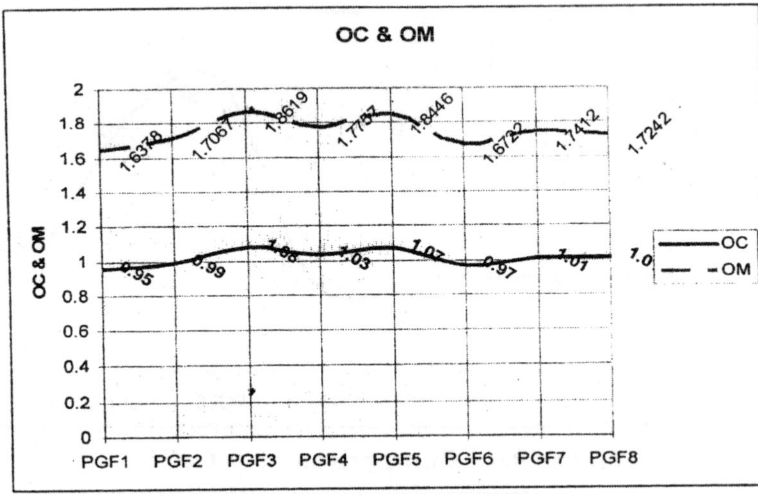

Graph 9.8 : Comparison of N1 and K

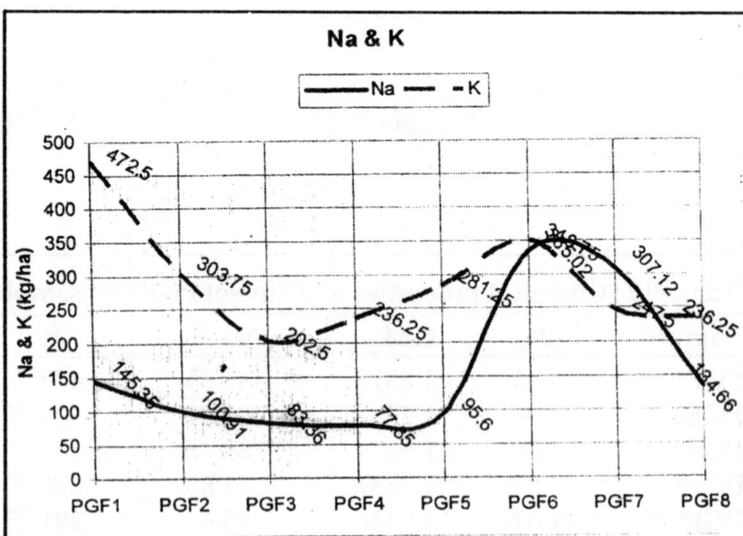

ANNEXUERE 1

SAMPLE CODE: PGF1

Name of Farmer	: Jaganath Raundal
Address of field	: Sector no. 236-237 Daregaon, Taluka Malegaon, District Nasik
Cultivated area	: 2 acres
Type of cultivation	: Monoculture
Varity of plant	: Syndreea Murdolla
Age of plant	: 6 years
Type of Irrigation	: Drip irrigation
Type of fertilizer used	: Bio-fertilizer

SAMPLE CODE: PGF2

Name of Farmer	: Jaganath Raundal
Address of field	: Sector no. 236-237 Daregaon, Taluka Malegaon, District Nasik
Cultivated area	: 2 acres
Type of cultivation	: Monoculture
Varity of plant	: Arrakta
Age of plant	: 5 years
Type of Irrigation	: Drip irrigation
Type of fertilizer used	: Bio-fertilizer

SAMPLE CODE: PGF3

Name of Farmer	: Shivaji More
Address of field	: Daregaon, Taluka Malegaon, District Nasik
Cultivated area	: 2 acres
Type of cultivation	: Monoculture
Varity of plant	: Syndreea
Age of plant	: 2 years
Type of Irrigation	: Drip irrigation
Type of fertilizer used	: Bio-fertilizer

SAMPLE CODE: PGF4

Name of Farmer	:	Shivaj i More
Address of field	:	Daregaon, Taluka Malegaon, District Nasik
Cultivated area	:	2 acres
Type of cultivation	:	Monoculture
Varity of plant	:	Syndreea
Age of plant	:	5 years
Type of Irrigation	:	Drip irrigation
Type of fertilizer used	:	Bio-fertilizer

SAMPLE CODE: PGF5

Name of Farmer	:	Ashok Patil
Address of field	:	Malda,Taluka Malegaon, District Nasik
Cultivated area	:	3 acres
Type of cultivation	:	Monoculture
Varity of plant	:	Arrakta
Age of plant	:	6 years
Type of Irrigation	:	Drip irrigation
Type of fertilizer used	:	Bio-fertilizer

SAMPLE CODE: PGF6

Name of Farmer	:	Kashinath Pawar
Address of field	:	Dayana,Taluka Malegaon, District Nasik
Cultivated area	:	2 acres
Type of cultivation	:	Monoculture
Varity of plant	:	Ganesh
Age of plant	:	6 years
Type of Irrigation	:	Drip irrigation
Type of fertilizer used	:	Bio-fertilizer

SAMPLE CODE: PGF7

Name of Farmer	:	Subhash Nikam
Address of field	:	Syoengaon near power sub-station Taluka Malegaon, District Nasik
Cultivated area	:	2 acres
Type of cultivation	:	Mixed Plantation
Varity of plant	:	Murdolla
Age of plant	:	4 years
Type of Irrigation	:	Drip irrigation
Type of fertilizer used	:	Bio- fertilizer

SAMPLE CODE: PGF8

Name of Farmer	:	Subhash Nikam
Address of field	:	Syoengaon near power sub-station Taluka Malegaon, District Nasik.
Cultivated area	:	3 acres
Type of cultivation	:	Mixed Plantation
Varity of plant	:	Ganesh
Age of plant	:	6 years
Type of Irrigation	:	Drip irrigation
Type of fertilizer used	:	Bio-fertilizer

ANNEXURE 2

Percentage value of N, P, and K present in different chemical fertilizers

Name of Fertilizers	% Nitrogen Oxide	% Phosphate Oxide	% Potassium
Nitrogen containing Fertilizers			
1. Ammonium Sulphate	20.6	---------------	---------------
2. Calcium Ammonium Nitrate	25	---------------	---------------
3. Ammonium Sulphate nitrate	26	---------------	---------------
4. Urea	46	---------------	
Phosphate containing Fertilizers			
1. Single Super Phosphate	---------------	16	---------------
2. Tripel Super Phosphate.	---------------	42.5	---------------
3. Dicalcium Phosphate	---------------	26	---------------
Potassium containing Fertilizers			
1. Murate of Potassh	---------------	---------------	58
2. Sulphates of Potassh	---------------	---------------	48
1. Nitro Phosphate	20	20	0
2. Nitro Phosphate (with Potash)	15	15	15
3. Mono Amrnonium Phospate	11	55	0
4. Di Ammonium Posphate	18	43	0
Mixed Fertilizers	34	17	0
	16	16	16
	10	36	10
	12	61	0
	11	19	19
	20	10	10

ANNEXURE 3
Percentage value of N, P, and K present in different bio-fertilizers

Sources of biofertilizers	(%)Nitrogen	(%)Phosphate Oxide	(%)Potassium Oxide
Compost	0.80	0.65	1.00
Goat dung	0.60	50	0.70
Poultry	3.03	2.36	1.40
Cow dung	0.56	0.35	0.78
Human excreta	1.30	1.10	0.35
Groundnutcake	7.10	1.40	1.30
Cotton seed cake	6.40	2.80	2.50
Caster seed cake	4.50	1.70	0.70
Bone crush	3.50	21.50	0.000
Fish vestiges	4.10	0.90	0.30

10

Distribution Pattern of aquatic plants in water-logged rice fields of Igatpuri Taluka of Nashik District, Maharashtra

Dilip Ahire

ABSTRACT

The water logged rice fields of the Igatpuri from Nasik district contains as wide variety of phytoplankton's. The distribution pattern of angiosperms, pteridophytes and algal flora of rice fields of Igatpuri taluka of Nasik District, revealed the abundance of 12 angiosperm, 3 species of pteridophytes and 4 species of algal flora.

Keywords : Aquatic Plants, Rice field, Phytoplankton.

Introduction

Igatpuri is small taluka of Nasik District. It lies between 20N latitude and 74 E longitudes of 599 m above main sea level. The climate is cool throughout the year. Vaitarna and Darana are two main rivers. The average annual rainfall is 10345 mm.

Water is the integral constituents of all the forms in the earth. A multitude of life forms are totally dependent on water for their growth, development and other metabolic activities. The plant community of water bodies like canals, lakes, rivers and even agricultural fields. In some parts of the world, aquatic plants in the form of weeds, are spreading at an alarming rate, and has become a serious global concern of financial losses to the farmers, traders and other departments as a result of their rapidly increasing population further, there emerges a competition between the crops grown in area and the phytoplanktons, resulting in considerable loss in the yield of crops. However, some of the phytoplanktons have tremendous economic potentiality since they influence the water environment as well as improvement of soil fertility of cultivated lands. (Venkataraman, 1969; Pandey et.al. 2001) Phytoplanktons are also of considerable ecological significance as they form the basic link in the food chain of all aquatic animals. (Mishra et.al. 2001). Rice is the principle food crops of Indian people, and like other crops, rice fields are also infested by a number of weeds in the growth and yield of rice crop has been reported by (Manivasagaperumal et.al. 2005). The present study aims at a preliminary survey of hydrobiology of water bodies in the rice fields of Igatpuri Taluka of Nasik district.

Material and Methods

The investigation was made in low land and rice fields of Igatpuri Taluka where the fields remain submerged almost throughout the crop growing season. For the qualitative estimation of phytoplanktons, water samples were collected as per procedure described by Hedge and Bharathi (1985) and Singh et.al.(1998) with the help of plankton net of mesh size 50 micron during the month from February 2008 - January 2009. The angiospermic flora were identified with the help of flora of the Bombay Presidency by Cooke, Flora of the Nasik district by Lakshminarasimhan and pteridophytes and algal flora at Department of Botany, Pune University.

Result and Discussion

A great diversity of aquatic plants are found to be available in the water in the rice fields of the Igatpuri taluka, a major rice growing area of the Nashik district, Maharashtra. The present survey was mainly concerned with the distribution pattern of Angiosperm, pheridophytes and algal flora (Table 1). A total 12 species of Angiosperm, were identified to be available in the fields almost throughout the year. Eichhornia crassipes, Lemna minor, Sagittaria sagittifolia, Limnophila indica, Eleocharis spirilis were found to be available to a maximum proportion during rainy season starting from May to August. When rice fields were totally submerged in water. Cyprus pilosus, Trapa bispinosa were present during the winter and automn season, while *Wolffia arrhiza* was present during the autumn season of August to November. Sphenoclea zeylanica was available in the water during the period from April to November. Among the pteridophytes, species of *Azolla pinnata* and *Salvinia cucullata* were available in the rice field during the month from June to August. While *Marsiela quadrifolia* were available during the winter season.

A variety of different algal flora, majority belong to the class chlorophyceae were isolated from the rice fields species of *Spirogyra* and *Oedogonium* were available during the rainy season starting from the water during the month of July. Species of *Nostoc* and *Anaebaena* belonging to cyanophyceae were isolated throughout the year except June and July.

Table 10.1. Distribution Pattern of aquatic plants in water-logged rice fields of Igatpuri Taluka

S.N.	Phytoplankton	Families	Phenology
1	**Angiosperm's:**		
	1. Ipomoea aquatica Forsk.	Convolvulaceae	Feb-Nov.
	2. Eichomia cressipes. Solm.	Pontederiaceae	Feb- Sept.
	3. Lemna minor Linn.	Araceae	June-August
	4. Limnophila indica. R. Br.	Scrophulariaceae	June-Sept.
	5. Pistia stratiotes Linn.	Araceae	June-August
	6. Sagittaria sagittifolia Linn.	Alismaceae	July-August
	7. Eleocharis spirilis Br. Prodr.	Cyperaceae	June-Oct.
	8. Trapa bispinosa Linn.	Trapaceae	Dec.-April
	9. Wolffia arrhiza Wimm.	Lemnaceae	Aug.-Nov.
	10. Cyperus pilosus Vahl. Enum.	Cyperaceae	Oct.-May
	11. Typha latifolia Linn.	Typhaceae	April-Aug.
	12. Sphenoclea zeylanica Gaertn.	Sphenoceaceae	April-Nov.
2	**Pteridophytes:**	**Class**	
	1. Marsilea quadrifolia L.	Filicinae	Dec.-March
	2. Azolla pinnata	Filicinae	June-Sept.
	3. Salvinia cucullata	Lycopodineae	June-Aug.
3	**Algae:**	**Class**	
	I. Spirogyra sp.	Chlorophyceae	April-Oct.
	2. Anabaena sp.	Cynophyceae	Sept.-April
	3. Nostoc sp.	Cynophyceae	Sept.-April
	4. Oedogonium sp.	Chlorophyceae	July-Aug.

Acknowledgement

Author is thankful to General Secretary, M.G. Vidyamandir and Principal M. S. G. College, Malegaon Camp for providing necessary facilities.

References

Cooke, T.C.I.E. (1967). The flora of the Presidency of Bombay.Vol. I, II and III. B. S. I. Calcutta.

Hedge, G.R. and Bharati, S.G. (1985). Comparative Phytoplankton of fresh water ponds and lakes of Dharwad, Kamataka State, India. Proc. Nat. Symp. Pure and Appl. Limnology.Bull. Bot. Soc. Sagar, 32:24-29.

Lakshinarasimhan, P. and Sharma, B.D. (1991). Flora of Nashik district.

Manivasagaperumal, R., Vijayarengan, P. and Vijayaravan, M. (2005). Effect of herbicides on biochemical changes in *Oriza* Sativa Linn. A cereal crop and Eclipta Alba Hass, a weed Ad. Plant Sd. 18(1 l):717-720.

Mishra, S.M., Pani, S., Bajpai, A., and Bajpai, A. K. (2001). Assessment of tropic status by using *Nygaard* index with references to Bhoj. Wetland. Pollution Research 20(2): 147-153.

Shah, G.L. (1978). Flora of Gujarat, Vol. I and II S.P. University, Vallabh, Vidyamandir.

Singh, S.N., Gupta, M.L. and Prakash, P. (1998). Hydrobiological studies of some eutrophic ponds of Rohtak, Bihar, Enviroment and Ecology.16(2):457-462.

Venkataraman, G.S. and Goyal, S. K. (1969). Influence of blue green algae on high yielding Paddy Variety IR-8. Sci. and Culture, 35-58.

PART II

Toxicological Studies

11

Toxicity of Fenevalerate to Fresh water Prawn *Macrobrachium lammeri*

B.S. Yadav

ABSTRACT

Experiments were designed to study the toxic effect of fenevalerate pesticide on fresh water prawn *Macrobrachium lammeri*. The LC 50 values for 24, 48, 72 and 96 hrs exposure periods framed out by using the finneys probit analysis method.

In the present probe it was observed that the LC 50 values for 24, 48, 72 and 96 hours exposure periods were 0.021 ppm, 0.018 ppm, 0.013 ppm and 0.008 ppm respectively. The highest LC 50 values were observed at 24 hrs exposure periods and it is found to decrease constantly with increasing the exposure period. The present result indicate that even low concentration of pesticide is also much toxic to fresh water prawn *Macrobrachium lammeri*.

Keywords : Toxicity, Pesticide, fenevalerate, LC 50 values, *Macrobrachium lammeri*

Introduction

Pesticide are known to pollute the environment through indiscriminate discharge of untreated or partially treated industrial effluents into water resources. It ultimately results in physical, chemical and biological changes which subsequently have direct and indirect effect on the aquatic fauna. Synthetic pesticide like fenevalerate constitute a serious type of pollution in fresh water fauna because they are stable compounds and are not readily removed by oxidation precipitation or by other means and affect the activity of the animals. Much of the work on crustaceans has been confined to the laboratory studies assessing direct toxicity of commonly occurring pollutants under various environmental conditions (Ahsanullah et. al. 1981, Murthi and Shukla 1984, Nagabhushanam et. al. 1986 and Sarogini et. al. 1990). Effects of pesticide on fresh water crustaceans are studied by few workers and very little is known about the effects of synthetic pesticide fenevalerate on fresh water crustaceans. Hence, present probe was undertaken to study the effect of fenevalerate pesticide on fresh water prawn *Macrobrachium lammeri*.

Materials and Method

The prawns *Macrobrachium lammeri* were procured from Girna Dam near Malegaon Dist. Nasik (Maharashtra State) India and acclimatised to the laboratory condition in the plastic troughs for 4-5 days prior to the commencement of the experiment. The prawns were fed to the pieces of earthworm and water was changed every alternate day. Approximately same size of prawns were used during the experiment.

The stock solution of fenevalerate pesticide was prepared by dissolving 1 ml pesticide in 500 ml of distilled water and from the stock solution different concentration were prepared.

Prawns were divided into several batches, each batch comprising of 10 prawns. Each batch of prawns were kept in plastic troughs containing two litter of water. The prawns were exposed separately to different concentrations of fenevalerate pesticide and mortality rate was recorded after 24, 48, 72 and 96 hours of exposure. The LC 50 values were calculated for 24, 48, 72 and 96 hours by using the Finney's probit analysis chart. (Finney 1971) Water parameters such as temp. pH, hardncss, dissolved oxygen, alkalinity, etc., were recorded.

Results and discussion

The results of the present probe clearly indicate that water used for the experiment did not contain any toxic substance and no mortality was observed in control group. Table 10.1 indicate the values of water parameters. Prawns were exposed to acute concentrations (LC 50 values of 24 hrs to 96 hrs). The LC 50 values are exprecessed in turm of ppm.

Table 10.1 : Shows the water parameters (values except PH and temperature are expressed in ml/L)

Parameter	Values
Temperature	26.5°C
pH	6.4
Dissolved Oxygen	4.64 ml
Alkalinity as CO_2	62 ml
Hardness	160 m1

Table 10.2: Relative toxicity of fenevalerate pesticide to prawn *Macrobrachium lammeri*.

Exposure Period (Hr)	LC 50±SE
24	0.021
48	0.018
72	0.013
96	0.008

Further as concentration increased there was increase in mortality. The LC 50 values decreased with increasing the exposure period. The values are highest at 24 hours and lowest at 96 hours of exposure. It is clear from the present results that percentage mortality increases with the increase in concentration of pesticide and exposure periods. Further it is observed that toxicity of a toxicant depends on many factors such as sex, size, and weight of the prawns. Sarojini et. al. 1990,development stages, Kamaldeep and Toor 1975, duration of exposure pH and hardness of water. (Handerson et. al. 1959) and dissolved oxygen content of the medium (Ester 1970). The synthetic pesticide fenevalerate in general affect the behaviour of the exposed prawns.

Chinnayya B. (1971) had noted the LC 50 value of heavy metals on shrimp *caridina rajadhari* Machele et. al. (1991) studied the effect of cuprous oxide on the fresh water crab. *Barytelphusa gulerini.* (Varghese et. al. (1992) reported the toxic effect of mercury, copper and zinc on fresh water crab *Barytelphusa cunicularis* (Shaikh I. S. 1996) studied the toxic effect of mercuric chloride on freshwater crab. *Barytelphusa guerini*. Pardeshi et. al. (2007) studied the Bio accumulation of mercury in different tissues of *Parreysia cylindrica* after acute and chronic exposure to mercuric chloride.

The results of the present investigation may be useful in evaluating the toxicity of pesticide fenevalerate to marine and fresh water crustaceans which from an important food chain of the aquatic ecosystem.

Acknowledgement

The author is very much thankful to the BCUD of Pune University authorities for providing the financial assistance.

References

Ahsanullah, M., D.S. Negilski and M.C. Mobley (1981) Toxicity of zinc, cadmium and copper to the shrimp, *Callianassa australiensis*. I Effects of individual metals. Mar, *Bid.* 64: 299-304.

Chinnayya, B. (1971) Effect of heavy metals on the oxygen consumption of the shrimp *Caridina rajadhari. Ind J. Exp. Biol.* 9(2) 277-278.

Eisler, R. (1970) Acute toxicity of organo chlorine and organophosphate insecticide to estuarine teleost. *Bull sport fish and wild Fish.* Tech, paper No. 46.

Finney, D. J. (1971) *Probit analysis,* 3rd ed. Cambridge University Press, Londan.

Handerson C., H.Q. Pickering and C.N. Tarzwell (1959) Relative toxicity of 10 chlorinated hydrocarbon insecticide to four species of fish. *Trans Amen. Fish Soc.* 88, 23-32.

Kamaldeep K. and H.S. Toor, (1975). Toxicity of pesticide to embryonic stages of Cyprinus. carpio (Linn) *Indian, J. Exp. Biol.* 15:193-96.

Machale, P.R., A.K. Khan, R. Sarojini, and R. Nagabhushanam (1991) Biochemical changes induced by cuprous oxide in the various tissues of the fresh water crab *Barytelphusa guerini. J. Adv. Zool.* 12(2) 115-20.

Murthi. R. and G.S. Shukla (1984) Toxicity of copper sulphate and Zinc sulphate to *Macrobrachium lamerrii* (H. Milne Edwords) Crustaceana 47(2) 168-73.

Nagabhushanam, R. K. Sambashiva Rao and R. Sarojini. (1986) acute toxicity of three heavy metals to marine edible crab *Scylla serrata. J. Zool.* 7, 97-99.

Pardeshi, Anilkumar and S.P. Zambare (2007) Bio-accumulation of mercury in different tissues of *Parreysia cylindrica* after acute and chronic exposure to mercuric chloride. *J. Aqua. Biol.* 22(1) 193-97.

Sarojini R, P.R. Machale, A.K. Khan and R. Nagabhushanam (1990) Effect of cadmium chloride on Histology and biochemical contents of the hepatopancreas of the fresh water crab *Barytelphusa guerini. J. Env. Series* (IV) 91-97.

Shaikh I.S. (1996) Toxic effects of heavy metals on some physiological aspects of crab *Barytelphusa guerini.* Ph.D. Thesis, submitted to Marathwada University Aurangabad, India.

12

Impact of Heavy Metal on Glycogen Content of Fresh Water Fish *Barbus ticto*

M.D. Nikam
B.S. Yadav

ABSTRACT

Impact of heavy metal Mercuric chloride on glycogen content of the liver and muscles in fresh water fish *Barbus ticto* were studied after acute exposure. The fishes were exposed to the lethal concentration of 1.95 ppm. 1.62, 1.40 and 1.20 ppm, LC50 values of 24 hrs. 48,72 and 96 hours respectively. After exposure period tissues like liver and muscles were pooled for biochemical assay.

The result of the present probe shows a significant decline in glycogen content of fish *Barbus ticto*. The decline in glycogen contents may be due to glycogenolysis in all tissues exposed to heavy metals.

Introduction

Fishes are recognized as valuable foods source and form important diet of man and has great nutritious value. But since several years production of fishes has been reduced due to contamination of aquatic ecosystem. Rives are exposed constantly to the newly developed industrial wastes, which includes heavy metals and oils. Heavy metals problem has been increasing day by day mainly due to industrial evolution, most of work on heavy metals pollution has been done on crustaceans and other species of fishes. Therefore present work has been undertaken to find out the impact of heavy metal. Mercuric chloride on fresh water fish *Barbus ticto*.

Materials and Methods

The fresh water fish *Barbus ticto* were collected from Girna Dam near Mategaon Nashik and reared in laboratory conditions for one week and pieces of earthworms were provided as a daily diet. Approximately same size and same weight of fishes were used for experiment. A batch of ten fishes were exposed to the LC 50 values of 24, 48, 72 and 96 hours. During the experiment fishes were not fed. Water parameters such as pH=7.2 temp. 27°C±1 Hardness 4.8 mg/L were measured. After exposure periods liver & muscles were dissected out and glycogen content were estimated by using the Kemp et. al. method.

Results

The results of the present investigation reveal that in liver and muscles a gradual decrease in glycogen contents were obtained after acute exposure to mercuric chloride with the concentration of 1.95, 1.62, 1.46 and 1.2 ppm (LC 50 value of 24, 48, 72 & 96 hours respectively).

Table-11.1 : Observation

Pollutant	Exposure period	Liver (mg/gm wet weight of tissues)	Muscle (mg/gm wet weight of tissues)
Control		3.2±0.026	1.8±0.032
Mercuric	24 hrs.	2.8±0.024	1.6±0.024
Chloride	48 hrs.	2.6±0.028	1.2±0.020
	72 hrs.	1.8±0.016	0.8±0.014
	96 hrs.	1.4±0.008	0.4±0.006

Discussion

Pollutants causes stress on the organism and disrupt its metabolic activities. The biochemical changes in the organs of animals exposed to pollutant have no definite pattern and the physiological state of metabolic activity of an organism reflects in the utilization of their biochemical energy to counteract the toxic stress (Nagabhushanam et. al.1987).

In the present experiment a significant gradual decrease in glycogen content were observed in liver of fresh water fish *Barbus ticto* after acute exposure to mercuric chloride. The decrease in glycogen content indicates its utilization for countering the heavy metal stress and it suggests prevalence of anaerobic conditions such as anoxia (Mayes 1977). Anoxia stress causes an increase in carbohydrate metabolison. This indicates that heavy metals mercuric chloride may cause stress conditions similar to anoxia and it also leads to increased glycogenolysis. Decrease in glycogen in the tissues may be due to higher energy demand during heavy metal exposure. Similar observation has been made by Jaiswal et. al. (1989) on fresh water prawn *Macrobrachium lamerrii* exposed to naphthalene. Nagabhushanam et. al. (1987) noticed a gradual decrease in glycogen content in hepatopancreas of fresh water prawns *M. Kistnensis* after exposure to pesticide theodon and phenthoate. Khillare et. al. (1987) reported a significant decline on tissue glycogen content of fresh water fish *Channa gachua* after dimethoate and Lindane exposure. Similar observation were made by Wam, et. al. (1984) on a fresh water fish *Garra mullya* after *thiourea* treatment. Patel (1991) reported the significant decline in glycogen content of fresh water crab *Barytelphusa cunicularis* exposed to endosulfan.

In the present probe when fresh water fishes were exposed to mercuric chloride a significant decline in muscles tissue were observed. The decrease in glycogen content may be due to enhancement of glycogenolysis by increase in phosphorelase activity similar observation were made by

Khan et. al. (1980) on fresh water crab *Barytelphusa guerinii* after exposure to zinc sulphate and copper sulphate Machel et. al. (1991) obtained a significant decrease in glycogen contents of crab *Barytelphusa guerinii* after acute exposure to cuprous oxide.

Above findings shows that pesticides after the biochemical sequence of metabolism and the animal is able to adapt to the stress condition.

Acknowledgement

Authors are thankful to the principal M.S.G. College Malegaon Camp. for providing the laboratory facilities.

Reference

Jaiswal, K.B., R. Nagabhushanam and R. Sarojini. (1989) Influence of Napthalene on the tissues of *Macrobrachium kistnensis* Geobios - 16(5) ; 197-202.

Kemp, A. and Van Heijmingem A.J.M. Kits. (1954) A Colorimetric micromethod for the determination of glycogen of fishes. Biochem. J. 56.

Khillare Y.K. and S.B. Wagh (1987) Effect of dimethoate and lindane on tissue glycogen content of the fresh water fish *Channa gachua. Proc Nat. symp. Ecotoxic* 91-95.

Khan, A.K., R. Sarojini, P.R. Machale and R. Nagabhushanam. (1990) Biochemical changes produced as a result of zinc sulphate and copper sulphate in the muscle of fresh water crab *Baritelphusa guerinii.* U.P.J. Zool 10 (1) 19-22.

Machale, P.R., A.K. Khan and R. Sarojini (1991) Biochemical changes induced by cuprous oxide in the various tissues of the fresh water *Crab B. gureinii. J. Adv.* 2001. 12(2) 115-20.

Mayes, P.A. (1977) Review of physiological chemistry. 16th. edn. Harper, H.A. Rodwell, U.V. and Mayes P.A. large Medical Publication California.

.Nagabhushanam, R., Jaya, Deshpande and R. Sarojini (1987). Effects of some pesticides on the biochemical constituents of the fresh water prawn. *Macrobrochium Kistnensis.*

Patel, R.I. (1991) Effect of endosulfan pesticide on same physiological aspects of *Barytelphusa guerinii*. Ph.D. Thesis. Marathawada University Aurangabad M.S.

Wam, G.P. and A.N. Latey. (1984) Effect of thiourea treatment on cholesterol contents of the liver and the Gonads of a freshwater teleost fish *Garra mullya*. *Geobios*, 11, 45-47, vol. II (1)

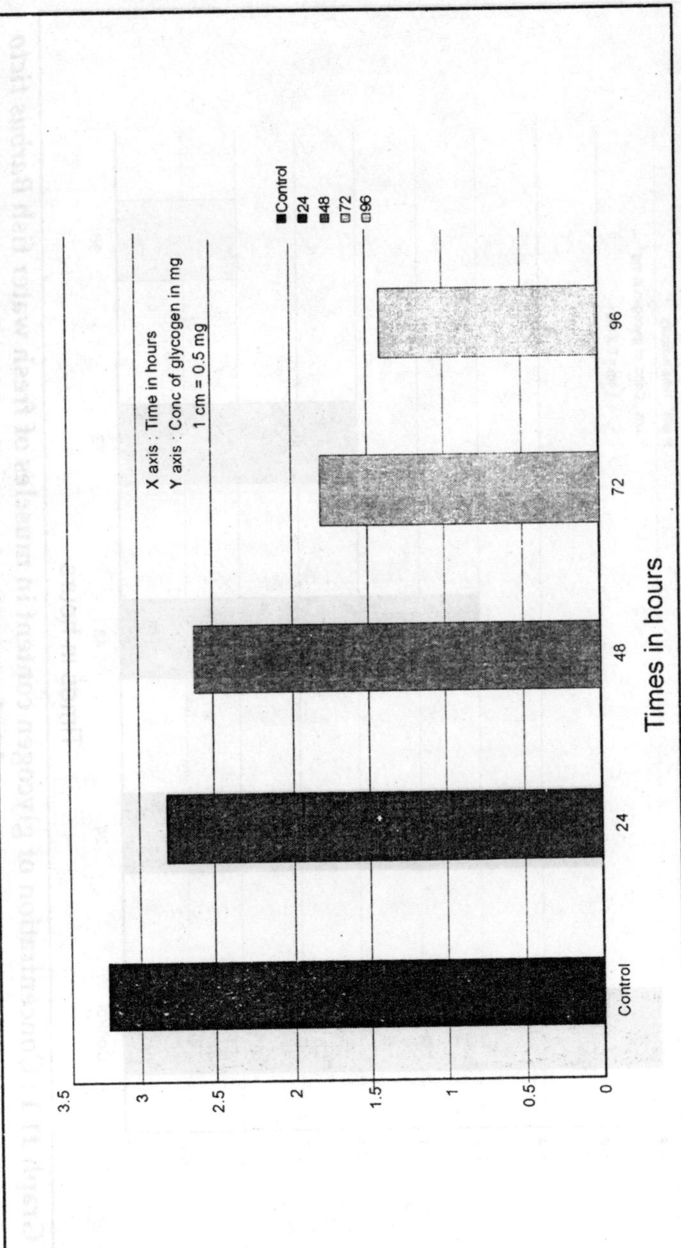

Graph 11.1 : Concentration of glycogen content in liver of fresh water fish *Barbus ticto* exposed to heavy metal nurcuric chloride

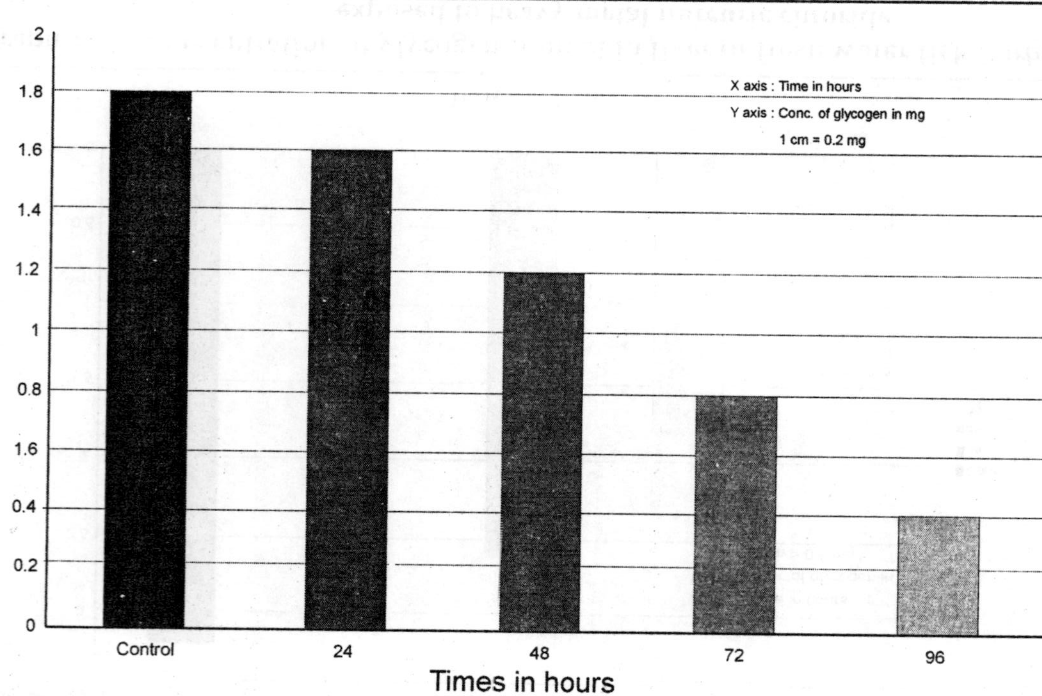

Graph 11.1 : Concentration of glycogen content in muscles of fresh water fish *Barbus ticto* exposed to heavy metal nurcuric chloride

13

Effect of Leafan-50 Pesticide on Erythrophores of the Fresh Water Crab *B. cunicularis*

M.J.Mahevi
B.S. Yadav

ABSTRACT

The colour changes in the crustaceans are accomplished by the dispersion of pigment granules in specific cells, the chromatophores. It is controlled by hormones produced by the eye stalk. The present investigation was carried out to determine whether Leafan-50 pesticide affect the colour changes of the fresh water crab *B. cunicularis*. The erythrophores of the crab exposed to different concentration of Leafan-50 were dispersed more when compared to control groups. The present experiment reveals that the dispersion of erythrophores are dose dependent and are controlled by the harmones of the central nervous system. Further the red pigments dispersing potency was less in the Leafan-50 exposed crabs when compared to the control group.

Introduction

The colour changes in the crustaceans are accomplished by the migration of pigment granules in specific cells called the chromatophores. In most of the crustaceans red, black and white chromatophores are present. Colour changes have been divided into two categories morphological and physiological. The morphological colour changes are generally evoked by keeping an animal on a specific background for a number of days. However physiological colour changes are evoked by certain environmental factors such as light and temperature.

Leafan-50 pesticide is commonly used by the former to increase the crop yield, and these pesticide finds way into the aquatic ecosystem through the run off of rain water and affects the Aquatic organisms. Crustaceans are one of the groups most sensitive to water pollution. Only a few studies have been conducted to determine the effects of environmental pollutants on the colour change mechanism in crustacean. Fingerman, et. al. (1978) noticed the less dispersion of chromatophores in fiddler crab *Uca pugilator* exposed to the polychlorinated biphenyl preparation.

Review of literature reveals that there is no information on the dispersion of chromatophores on crustaceans exposed to the pollutants, therefore present work has been undertaken to study the effect of Leafan-50 on colour physiology of the fresh water crab. *B. cunicularis.*

Materials and Methods

The fresh water crabs *Barytelphusa cunicularis* were collected from Girna Dame near Malegaon, Nasik and reared in plastic troughs containing aerated tap water. Crabs were fed alternate day with the pieces of earthworm and water was changed daily. The crabs were acclimatized in the laboratory condition for 4-5 days prior the commencement of the experiment. Crabs were not fed during the experiment. During the experimentation approximately same size of crabs were used. The crabs were used in each experiment without regard to *Sex*. The Black and Red chromatophores on the walking legs were observed under the microscope in both, control and pesticide exposed crabs.

The stock solution of Leafan-50 pesticide was prepared by dissolving 1 ml of pesticide in 500 ml. of tap water and from the stock solution different concentrations were prepared.

Results

The results of the present investigation reveals that when crabs were exposed to the 6 ppm of Leafan-50, the dispersion of red chromatophore induced in 3 hrs. However when the concentrations were increased to 8 ppm the dispersion was maximum after 3 hrs. exposure. The present experiment reveals that dispersion of red pigment was concentration dependent.

On the other hand when crabs were exposed to 15 ppm of Leafan-50, after 3 hrs. of exposure the red chromatophores showed less dispersion potency.

Discussion

In the present investigation it was observed that the dispersion of red chromatophores was dose dependent. When crabs were exposed to 6 ppm and 8 ppm of pesticide Leafan-50, their was maximum dispersion in red chromatophores. However when crabs were exposed to 15 ppm of pesticide there was reduction in red chromatophore dispersion. Similar observation were made by Nagabhushanam et. al. (1991) on fresh water prawn *Caridina rajadhari* exposed to Naphthalene up to certain extent of Naphthalene exposure, which may be due to increase in the amount of melanin dispersing hormone in the eye-stalk of the crab. Nagabhushanam et. al. (1988) reported similar dispersion of red chromatophores of fresh water prawn *Macrobrachium kistnensis* after exposure to pesticide sevimol. and suggested that sevimol is possibly acting through the *neurotransmitter* which control the melanin dispersing hormone. Fingerman et. al. (1981) studied the effect of various biogenic amines on colour changes of *Uca pugilator* and put forth evidence that 5-hydroxyamine and Dopamine act as Neurotransmitters triggering release of erythrophorotropic hormone. Presence of various biogenic amines in crustaceans nervous tissues have been reported by many workers (Butter and Fingerman, 1983, Laxmyr, 1984). In the present study Leafan-50 might have altered the concentrations of various biogenic amines such as 5-hydroxyamine and epinephrine.

Another possibility for the red pigment dispersing response to the Leafan-50 may be through epinephrine which is known to enhance the accumulation of intracellular cyclic AMP by activating a membrane bound enzyme adenylase cyclase which catalyses the conversion of the ATP to cyclic AMP. Cyclic AMP then initiates a series of intracellular alterations resulting in the characteristic metabolic effects of catecholamine. Vasantha (1968) observed the red pigment dispersing response to cyclic AMP in fresh water prawn *Caridina weberi*. Similar observations were made by Nagabhushanam et. al. (1991) on fresh water prawn *Caridina rajadhari* exposed to naphthalene. Sarogini, et. al. (1987) reported the similar results on fresh water prawn *Macrobrachium kistnensis* exposed to phytoalkaloids *Vinca rosea*.

Acknowledgement

The authors (MJM and BSY) are thankful to the principal, MSG College, Malegaon Camp for providing the laboratory facility to complete this work.

References

Butler T.A. and M. Fingerman. (1983) Concentrations of neurotransmitters in the central Nervous System of *Uca Panacea* and *Callinectis Sapidus*. *Am. Zool.* 23:954.

Fingerman, S.W. and M. Fingerman (1978) Influence of the polychlorinated biphenyl preparation Arachlor 1242 on colour changes of the fiddler crab. *Uca pugilator*. *Mar. Biol.* 50:37-45.

Fingerman, M., M.M. Hanumante and S.W. Fingerman. (1981) The effects of Biogenic amines in colour changes of the fiddler crab

Uca pugilator. Further evidence for role of 5-Hydroxytryptaminc and dopamine on Neurotransmitters triggering release of erythrophore tropic hormone comp. *Biochem, Physiol* 68. C. 205-11.

Laxmyr. L (1984) Biogenic amines and DOPA in the central nervous system of decapod crustaceans. Camp. *Biochem. Physiol.* 77 139-43.

Nagabhushanam, R., P.R.Machale and R. Sarojini (1988) Effect of sevimol on red chromatophor physiology of fresh water prawn *Macrobrachium kistnensis J.Zool. Res.* 1 (2) 101-03.

Nagabhushanam R., P.R. Machale, R.V. Katyani, P.S. Reddy and R. Sarojini (1991) Erythrophoretic responses induced by Naphthalene in fresh water prawn *Caridina rajadhari. J. Ecotoxicol. Environ. Monit.* (3) 185-91.

Sarojini, R., G.K. Kulkarni, A.K. Khan and R. Nagabhushanam (1987) Effect of phytoalkaloids from *Vinca rosea* on chromatophores of a fresh water prawn *Macrobrachium kistnensis. Proc. Nat. Symp. Ecotoxic.* pp. 8-12.

Vasantha, N. (1963) Some aspects of endocrinology of the fresh water prawn *Caridina weberi.* Ph.D. thesis, Marathwada University. Aurangabad. M.S. India.

14

Study of alteration in respiratory metabolism of fresh water crab *Barytelphusa cunicularis* after lethal exposure to "Confidor"

K.T. Patil
B.S. Yadav

ABSTRACT

Impact of confidor was carried out on the respiratory rate of fresh water crab *Barytelphusa cunicularis* after acute exposure. In acute condition the crab were exposed to 2.8 ppm, 2.3 ppm, 1.8 and 1.2 ppm, the LC50 values of 24, 48, 72 and 96 hrs, respectively. After exposure period the respiratory metabolism were studied. After acute exposurę an initial increase after 2, 4, 6 and 8 hours exposure and then significant decrease in level of oxygen consumption was observed after 12, 24, 48, 72 and 96 hours exposure periods of confidor pesticide.

Introduction

To understand the animal respiration we must consider not only the nature of respiratory organ but also the mechanisms used to control the respiration and the nature of respiratory systems adaptations to different environment. The poisoning symptoms of pesticide consumption during the initial stage of intoxication clearly suggest that their primary action is on the nervous system, due to which respiratory chains are inhibited.

Pesticides are well-known pollutants which cause disorders in aquatic ecosystem with deleterious effects on associated organisms (Kaviraj and Konar 1982). The rate of oxygen consumption is considered as a reflection of the total metabolism and metabolic rate of the aquatic organisms. Living organisms have the ability to adopt themselves to changes in the environmental parameter such as temperature, humidity, oxygen supply or a pesticide stress (Hoar 1976). The process of respiration is one of the best known physiological activities in the crustacean, therefore in the present investigation, confidor pesticides was used to see the toxic effect of the pesticide on respiratory metabolism of fresh water crab *Barytelphusa cunicularis*.

Material and Methods

The fresh water crab *Barytelphusa cunicularis* were collected from Girna Dam near Malegaon, Nasik district and acclimatized to the laboratory condition for 4-5 days in the plastic trough. The crabs were fed with peices of earthworm alternate days and water was changed. For the experiments approximately same size of crabs were used.

The water parameters such as pH =7.4, Temperature 27±2 and Hardness 230 mg/L were maintained.

The crabs were exposed with the LC50 value of confidor for 24, 48, 72 and 96 hrs. After exposure periods the oxygen consumption were measured after 1, 2 , 4, 6, 8,12, 24, 48, 72 and 96 hrs. by using the modified winklers methods.

Results

The results of the present investigation on respiratory metabolism of fresh water crab *Barytelphusa cunicularis* after confidor exposure are shown in Table 13.1 Figure 13.1. When crabs were exposed to 2.8 ppm (LC5o value of 24 hrs) a significant increase in oxygen consumption were observed up to 12 hours of exposure. However after 24 hrs a gradual decrease in rate of oxygen consumption were observed after 48, 72 and 96 hours exposure also shows a significant decrease in oxygen consumption.

Table 13.1 : Oxygen consumption of *Barytelphusa cunicularis* in different concentration of pesticide 'confider'

Exposure Periods → (In hrs)	2	4	6	8	12	24	48	72	96
Control	.0541	.0624	.0642	.0742	.0746	.0721	.0624	.0624 .	0664
±	.0046	.0034	.0042	±	±	±	±	±	
					.0005	0.0062	.0034	.0006	.0004
Experimental	0.0062.	.0746	.0768	.0864	.0842	.0646	.0526	.0486	.0424
	±.0004	±.0004	±.0056	.0058	±	±	±	±	±
					.0008	.0062	.0042	.0028	.0006

Discussion

Oxygen consumption of an animal is a very sensitive physiological process and changes in the respiratory activity have been used as indicator of stress in taxicant exposed animals. The measurements of oxygen consumption of animal therefore would provide an additional clue to the physiological mode of action of pesticide pollutants.

When crabs were exposed to the lethal concentration of pesticide "confidor" a significant increase in oxygen consumption were observed up to 8 hours of exposure. An initial increase in rate of oxygen consumption showed immediate response and onset of compensatory mechanism in order to overcome the harmful effects of pesticides.

Many workers have shown an initial increase in rate of oxygen consumption. Martin et. al. (1990) reported the similar results on fresh water prawn *Caridina weberi* exposed to methyl parathion. An initial increase in rate of oxygen

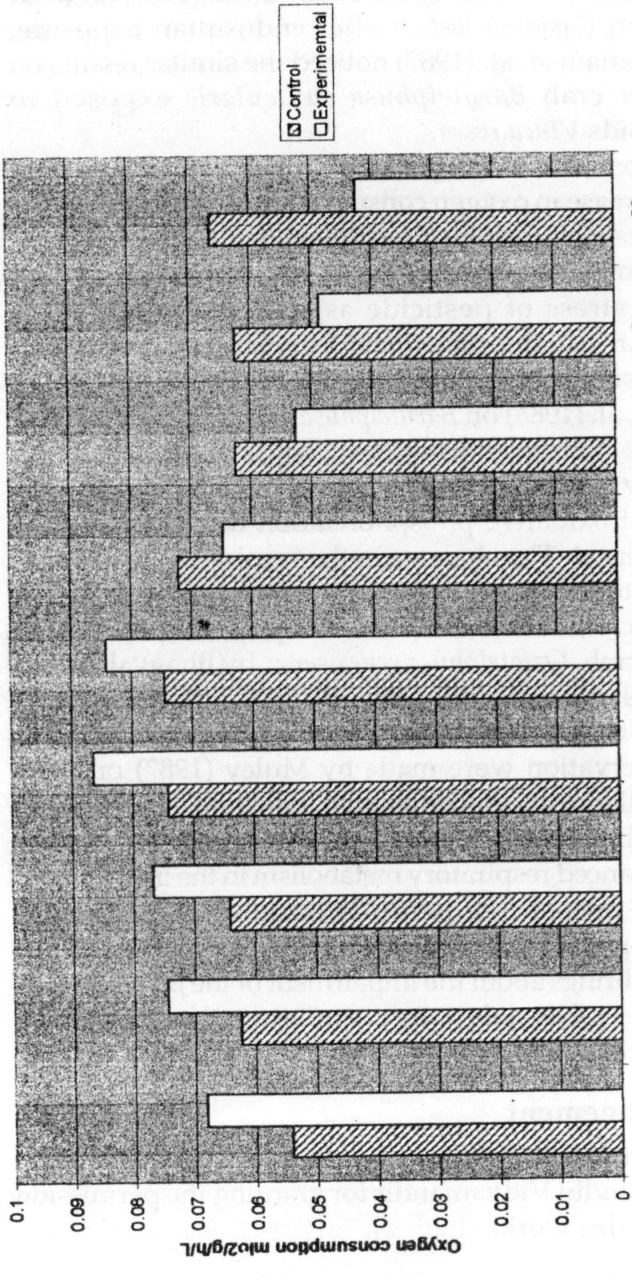

Graph 13.1 Oxygen consumption in fresh water crab *Barytelphusa cunicularis* after exposure of 'confidor'

consumption were observed by Yadav et. al. (1989) on fresh water prawn *Caridina weberi* after endosulfan exposure. Nagabhushanam et. al. (1987) noticed the similar results on fresh water crab *Barytelphusa cunicularis* exposed to phytoalkaloids *Vinca rosea.*

In the present investigation an initial increase and then gradual decrease in oxygen consumption was observed. The decrease in rate of oxygen consumption may be due to failure of crab to compensate for the new steady state of metabolism due to the stress of pesticide as the exposure periods increases. An initial increase and then gradual decrease in oxygen consumption in lethal concentration was observed by Bodke et. al. (1985) on *Barytelphusa cunicularis* exposed to sevimol. The decrease in rate of oxygen consumption of crab *B. cunicularis* exposed to confidor may be due to the inhibition of oxidative phosphorelation due to prolonged exposure period. The changes in the respiration of animals may be due to the impairment of certain key enzyme system of metabolic significance has been reported in sumithion exposed to crab. *Oziotelphusa senex senex* by Bhagyalakshmi (1981). Similar result were obtained by Khan et. al. (1937) when crab *Barytelphusa cunicularis* were exposed to Benzene. Similar observation were made by Muley (1987) on fresh water Lamillibranch on endosulfan exposure.

From the above results it can be concluded that crabs showed enhanced respiratory metabolism in the initial stages of pesticide impact. The pesticide confidor may affect the metabolic pathway. The changes in the respiratory metabolism brings about the impairment of the physiological fitness of the crab and there by contribute to the lethal action of the pesticide.

Acknowledgement

The authors are thankful to the co-ordinator of Mahatma Gandhi Vidyamandir for granting the permission to complete this work.

References

Bhagyalakshmi, A. (1981) Physiological studies on the fresh water field crab *Oziotelphusa senex-senex* in relation to pesticide stress. Ph.D. Thesis Shri. Venkateswara University Tirupati India.

Bodke, M.K., R. Nagabhushanam and R. Sarojini (1985) Effect of endocrine manipulation and carbamate exposure on the oxygen consumption of the fresh water crab *Barytelphusa cunicularis* comp. *Physiol. Ecology.* 10 (3) 105-03.

Hoar, W.S. (1976) *General and comparative physiology,* Printice Hall, New Delhi.

Kaviraj and S.K. Konar. (1982) Acute toxicity of mercury, chromium and cadmium to fish plankton and worms. *Geobios* 9 (3) 97-100.

Khan, AK, K. Sambashiva Rao, and R. Sarojini (1987). Effect of lethal and sublethal concentration of Benzene on the respiratory metabolism of the fresh water crab *Barytelphusa cunicularis. Nat symp. Phy. crusta.* 22-26.

Martin, E.R and R. Sarojini (1990) Sex associated changes in respiratory metabolism of fresh water prawn. *Caridina weberi* exposed to melthyl parathion. *J Adv.* zoo. 2001.1(2), 85- 88.

Muley, D. V. and U.H Mane (1987) Endosulfan induced changes in oxygen consumption in two species of fresh water lamellibranch molluscs. *Indian. J. Com Anim. Physiology,* vol.5, no.1, 28-33.

Nagabhushanam, R., R. Sarojini and A.D. Diwan. (1987) Oxygen consumption in the fresh water prawn *Macrobrachium kistnensis* and the crab *Barytelphusa cunicularis* as a function of the phyto alkaloids Vinca rosea.

Yadav, B.S. and R. Sarojini (1989) changes in Respiratorv metabolism of fresh water prawn *Caradina weberi* after exposure to endosulfan proc. Nat.symp.anim. prot. changing-8 env. Him J. Env. 200 vol. 3, 144-48.

15

Respiratory responses of the fresh water crab Barytelphusa guerini exposed to the pesticide Folicur (Fungicide)

B.S. Yadav
Homaira Badruzzama

ABSTRACT

The crabs were exposed to the lethal concentration of 19.95 ppm, 9.3 ppm, 7.2 ppm, 3.7 ppm of the folicur, LC5o values of the 24, 48, 72 and 96 hrs. respectively for acute and for subacute study the crabs were exposed to the 1/10 LC5o values of 48 hrs. i.e. 0.93 ppm for 10, 20 and 30 days. After exposure periods the oxygen consumption were calculated by using the modified winklers azide method. The results shows a significant decrease in oxygen consumption when exposure periods increases. The reduction of oxygen consumption may be due to reduction of glycolysis in order to overcome the anaerobic stress caused by the fungicide Folicure.

Keywords : Crab, Folicure, oxygen consumption, LC50 values.

Introduction

Aquatic organism respond quickly to pollutants by altering their metabolic rate. The measurement of oxygen consumption of the animal in the presence of a particular pollutant is the best index of its activity. In general both intrinsic and extrinsic factors can influence the respiratory capacity of the organism.

Alternation of oxygen consumption of whole organisms or isolated tissues by different pollutants or toxicants have been employed to measures the condition of an organisms (Vernberg and Veinburg 1972; Brown and Newell 1972; Thurnberg et. al. 1973; Delthay and Corner 1975; Umadevi 1987).

Many workers have reported the effects of copper and organophosphate pesticides on the oxygen consumption of aquatic animals, but meager literature is available on fresh water crabs. Therefore present work has been undertaken to study the lethal and sublethal impact of pesticides folicure on fresh water crab *B. guerini.*

Materials and Methods

The crab *Barytelphusa guerini* were collected from Girna Dam near Malegaon Dist. Nasik. Maharashtra and acclimatized to the laboratory condition for 4-5 days in the plastic troughs. Water was changed daily and crabs were fed alternate days with the pieces of earthworm. For the experiment approximately same size of the crabs were used.

In the present study for respiratory metabolism crabs were divided into five batches and each batches was exposed to different concentration of the pesticides.

The first batch of crabs was exposed to 19.95 ppm of folicur(LC50 value of 24 hrs.) Second, third and fourth batches were exposed to 9.3 , 7.2 and 3.7 ppm, Lc50 values of 48, 72 and 96 hours respectively. The fifth batch of the crabs were exposed to 0.93 ppm of folicure (1/10 LC 50 values of 48 hrs) for sublethal study.

After exposure periods oxygen consumption was measured by using the standard Wrinkler's method.

Results

From the present observation it is clear that when crabs were exposed to 19.95 ppm of folicure (LC50 value of 24 hrs) a significant decrease in rate of oxygen consumption was observed. Similarly gradual decrease in rate of oxygen consumption was observed in the crabs after exposure to the LC50 values of 48, 72 and 96 hrs.

At sublethal exposure the crabs showed a significant decrease in rate of oxygen consumption when compared to the control groups.

Table 14.1

Concentration of Pesticide (pp^m)	mg Oxygen/hr/gm. body wt/l
Control	0.379
19.95ppm 24 hrsLC50 values	0.232
9.3 ppm 48 hrs LC50 values	0.13 8
7.2 ppm 72 hrs LC50 values	0.114
3.7 ppm 96 hrs LC50 values	0.076
0.93 ppm **(fa 10 days)1/10 LC50 values)**	0.034
0.93 ppm **(fa 20 days)1/10 LC50 values)**	0.025
0.93 ppm **(fa 30 days)1/10 LC50 values)**	0.0 14

Table-14.1 shows the decrease in rate of oxygen consumption exposed to acute and chornic concentration of the pesticide folicur.

Graph 14.1 indicates the decrease in rate of oxygen consumption in fresh water crab *Barytelphusa guerini* exposed to acute and chronic concentration of the pesticide folicure.

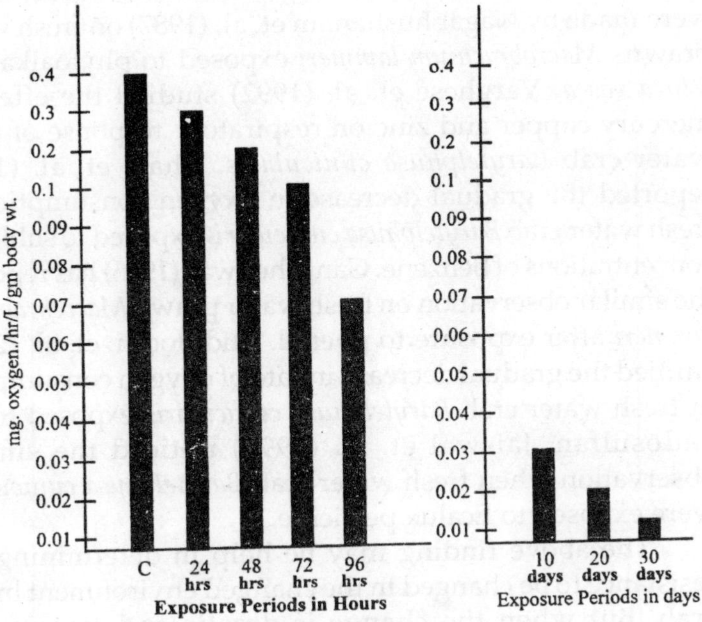

Hours and Days

Discussion

The rate of oxygen consumption of crab *Barytelphusa guerini* has been found to be diminished after exposure to the folicure. The result of the present investigation reveals a gradual decline in rate of oxygen consumption may be due to failure of crabs to compensate for the new steady state of metabolism due to the stress of folicure as the exposure period increased continuously. This findings were supported by many workers. Yadav, et. al. (1989) studied the impact of endosulfan on respiratory metabolism of fresh water prawns *Caridina weberi*. Further he reported that decrease in rate of oxygen consumption of prawn *Caridina weberi* exposed to endosulfan may be due to the inhibition of oxidative phosphorelation due to prolonged exposure periods. Thus in the present results folicure pesticide possibly may inhibit certain enzymatic pathways in the respiratory metabolism of fresh water crab *Barytelphusa guerini*. Similar observation were made by Nagabhushanam et. al. (1987) on fresh water prawns *Macrobrachium lammeri* exposed to phytoalkaloids *Vinca rosea*. Varghese et. al. (1992) studied the effect of mercury copper and zinc on respiratory response of fresh water crab *Barytelphusa cunicularis*. Khan, et. al. (1987) reported the gradual decrease in oxygen consumption of fresh water crab *Barytelphusa cunicularis* exposed to sublethal concentrations of benzene. Gangshetiwar (1986) has reported the similar observation on fresh water prawn *Macrobrachium lammeri* after exposure to phenol. Mididoddi, et. al. (2005) studied the gradual decrease in rate of oxygen consumption in fresh water crab *Barytelphusa cunicularis* exposed to the endosulfan. Jaiswal et. al. (1990) noticed the similar observation when fresh water crab *Barytelphusa cunicularis* were exposed to Ecalux pesticide.

The above finding may be help in determining the resistance to be changed in the changed environment by the crab. But when the change is drastic and toxic one it ultimately depresses the metabolism, causing damage to the respiratory epithelium and differing capacity of gills and thereby bringing a change in level of oxidative enzymes.

Acknowledgement

Authors are thankful to the principal of M.S.G. College for providing the laboratory facilities.

References

Gangashetiwar, C.V.B. (1986) Effect of phenol poisoning on the physiology of the Prawn *Macrobrachium lammeri*. Ph.D thesis Marathwada University Aurangabad M.S. India.

Jaiswal, K., B.D Deshmukh and R. Sarojini. (1990). Exposure to organophosphorus insecticide (Ecalux) alters the oxygen consumption in the fresh water crab *Barytelphusa cunicularis*.*U.P. J. Zool* 10(1)108 - 10.

Khan, A.K, K. Sambashivarao, R. Sarojini. (1981) Effect of lethal and sublethal concen-tration of Benzene on the respiration of the fresh water carb *Barytelphusa cunicularis*. Nat. symp. phy. crust. 22.26.

Mididoddi, Venkateshwaralu and Mollor Shanmugam (2005). Effect of endosulfan on oxygen consumption and succinate and lactate dehydrogenase activities in fresh water crab. *Barytelphusa cunicularis* Biologia, Bratishave 60, 3 07-311.

Nagabhushnam, R., R. Sarojani and A. K. Khan (1987) Oxygen consumption in fresh water prawn *Macrobrachium kistnensis* and the crab *Barytelphusa cunicularis* as function of the phytoalkaloids from *Vinca rosea*. *Proc. Nat. Symp. Ecotoxic* 39-41.

Thernberg, F.P., M.A. Dowson and R.S. Colliar (1973) Effects of Cu and Cd on osmoregulation and respiration in two species of estuarine crab. *Marine Biology*, 23, 171-75.

Varghese G., P.S. Naik, M. Katdare (1992) Respiratory responces and blood sugar level of the crab *Barytelphusa cunicularis*. exposed to mercury, copper and zinc. *India.J.Exp.Biol.* 30(4)308-12.

Vernberg W.B, and F.J. Vernberg (1972) The synergistic effects of temperature, salinity and mercury on survival and Metabolism of the adult fiddler crab. *Uca pugilatar*. *Fish. Bull.* 70, 415- 20.

Yadav, B.S., R. Sarojini and R.Nagabhushanam (1989) Changes in respiratory metabolism of fresh water prawn *Caridina weberi* after exposure to endosulfan. proced. *Nat. Symp. Anim. Proc. changing Envi.* Pulished by *Him J. Envi Zool*, vol. 3, 144-48.

16

Seasonal evaluation of toxicity levels of Monocrotophos on mortality and behavioural responses of local fresh water Fish, *Nemacheilus botia*

J.D. Vasait
V.T. Patil

ABSTRACT

The acute toxicity of Monocrotophos on the mortality and behaviour of *Nemacheilus botia* has been studied. The LC50 values for 14 days exposures are 64.5, 42 and 31.3 mg/lit, during winter, summer and rainy seasons respectively. The observed result indicates that the mortality of the test fish to Monocrotophos was dose-time dependant. The altered behavioural responses were also observed under investigations in the test organism exposed to Monocrotophos.

Keywords: Seasonal, Bioassay, Monocrotophos, Mortality, Behaviour, *Nemacheilus botia*.

Introduction

Toxicity studies have long played an important role in man's efforts to monitor and modify the effects of his activities on the biota. The toxicity studies especially useful in determining the sensitive species of an ecosystem that can be used as indicator species, for a particular type of pollution. The results of toxicity are generally reported in terms of median lethal concentration LC50 or median tolerance limit (MTL).

The major groups of insecticides in common use are the organophosphorous compounds. Many of the organophosphates are considered hazardous because of their ability to kill or immobilize various organisms at extreme low conditions (Eisler, 1969). Organophosphates are highly toxic to fish and are powerful nerve poisons, since they inhibit AchE activity (Casida, 1964; Coppage and Mathews, 1975; Klavercamp and Hobdon, 1980 and Rath and Misra, 1981).

Several workers have investigated the toxicity of organophosphorous pesticides in fish and other animals (Lockhart, et. al., 1973; Clavercamp, et. al., 1977; Johansson and Finley, 1980; Koundinya and Rammurthi, 1980; Devi, 1981; Srivastava and Singh, 1981;Murthi et. al., 1983; Sahib et. al., 1984; Prasada Rai and Ramana, Rao 1984; Patil, 1986; Chanchal, et. al., 1990; Vasanthi, et. al., 1990; Patli; 1991; Debeer and Rajyasree, 1934; M. Santhakumar, M. Bajaj and K. Ramadu 2000; Pramod Joshi 2002 and Subathra and Karup-pasamy, 2003).

The use of behavioural parameters in toxicological assessment of aquatic pollutant has been rare (Eisler, 1977). Behavioural changes are easily detectable as compared to complicated and complex processes underlying them. With this view behavioural changes are observed along with toxicity bioassay. A. Murugan (1992) observed the behaviour of fish *Labeo rohita* (Hamilton) on exposure of distillery effluents. Subathra and Karuppasamy (2003) observed acute toxicity levels of cadmium on mortality and behavioural responses of an air breathing fish, *Channa punctatus*.

Material and Methods

Nemacheilus botia is an inhabitant of Panzan river

flowing near Manmad (Nasik), India. The live fishes were netted from this river, brought to the laboratory and released in glass aquaria (size 3×1×1× ft.). The fishes were fed on fish meal procured from market and allowed to acclimatize to laboratory conditions for one week.

The physico-chemical characteristics of test water were analysed during experimentation period (APHA, 1989). The bioassay methods adopted in present investigation were the same as that of Doudoroff et. al. (1951). Fish of uniform size (3 $^{+}$1cm.) and weight (1^{+} 0.2 gm.) was selected for LC50 determination. An organophosphorous insecticide Monocrotophos 36 per cent W/w commercial grade procured from market and used for treatment. Preliminary experiments using different concentrations of Monocrotophos was conducted to find concentrations that resulted in 10 per cent -100 per cent mortality. After the range finding test, the concentrations of Monocrotophos 50 ppm to 85 ppm, 30 ppm to 60 ppm and 20 ppm to 50 ppm were selected for winter, summer and rainy seasons, respectively.

For experimentation, the laboratory-acclimatized fishes were exposed to the different concentrations of Monocrotophos. The water of aquarium was aerated twice a day to prevent hypoxic conditions and changed at every 24 hours. A batch of control fish was also maintained along with experimented fish.

The behaviour of the fish *Nemacheilus botia* was studied in the control groups as well as after exposure to Monocrotophos.

Results and Observations:

The LC50 values of Monocrotophos found for fourteen days exposure are 64.5, 42 and 31.3 ppm during winter, summer and rainy seasons, respectively. (Tables 15.1, 15.2, 15.3, and Graphs 15.1, 15.2, 15.3).

Animal behaviour (Control): The *Nemacheilus botia* shows shoaling behaviour. The constant opercular movement, normal body balance and posture were observed in control fish.

Treatment: The exposed fishes had shown the altered behaviours as compared to the control fish. The fishes were faced suffocation, irregular, erratic and jerky movements. The paralysis of body parts with unusual inclined posture also observed in fishes.

Discussion: In the present investigation the LcC0 concentrations of Monocrotophos were calculated for winter, summer and rainy seasons. It was found that; Monocrotophos is hazardous in every season.

During winter season the insecticide Monocrotophos was less toxic to the fishes and requires more concentration for LC50 dose. During rainy season Monocrotophos was found more toxic to the fishes even when that during summer season. Thus toxicity was found to be more during rainy season somewhat less toxic during summer season and least toxic during winter season. Bhatlawande (1989) observed, the LC50 value of Monocrotophos on the terrestrial snail, *Cerastus monssonianus* in three different reproductive periods and found variable concentrations required to LC50. Patil V.T. (1991) calculated the LC50 values of same insecticide Monocrotophos on the snail, *Indoplanorbis exustus* in three different periods. He has concluded that, the toxicity doses required variable in different periods. The above observations support the present investigation. However, variation of LC50 values may be due to changes in environmental factors or biology of fish.

The behaviour of fish is the important part of identification of stress conditions. The behaviour is the most significant activity of fish to identifying by the toxicity of fish.

In the present investigation the altered behaviour observed in the exposed fishes. The same behaviour is observed in the exposed fishes. The altered behaviour is observed by several workers in response to toxic substances in fishes (Panigrahi and Mishra, 1978; Malya Revya, 1978; Watenpaugh and Beitinger, 1985; Sen et. al., 1991; Arumugam, 1992; Amudha et. al., 1997; Barnah et. al., 2002 and Subathra S., 2003).

Acknowledgement

Author thanks to Hon'ble Prashantdada Hiray General Secretary, Mahatma Gandhi Vidyamandir, Malegaon camp and Dr. K.B. Patil, Vice-chancellor, N.M. university, Jalgaon for providing necessary laboratory facilities.

Table 15.1 : Mortality of fish, Nemacheilus botia at different concentrations of monocrotophos for 14 days exposure period in winter season

Conc. of Monocro-tophos (ppm)	Log of Conc.	Fish Exposed	Fish dead	% Kill	Probit kill
50	1.70	10	01	10	3.71
55	1.74	10	01	10	3.71
60	1.78	10	03	30	4.47
65	1.81	10	05	50	5
70	1.85	10	07	70	5.52
75	1.88	10	08	80	5.84
80	1.90	10	09	90	6.28
85	1.93	10	10	100	8.09

Table-15.2 : Mortality of fish, *Nemacheilus botia* at different concentrations of monocrotophos for 14 days exposure period in summer season.

Conc. of Monocro-tophos (ppm)	Log of Conc.	Fish exposed	Fish dead	% Kill	Probit kill
30	1.48	10	01	10	3.71
35	1.54	10	03	30	4.47
40	1.60	10	04	40	4.74
45	1.65	10	05	50	5
50	1.70	10	06	60	5.25
55	1.74	10	08	80	5.84
60	1.78	10	10	100	8.09

Table 15.3 : Mortality of fish, Nemacheilus botia at different concentrations of monocrotophos for 7 days exposure period in rainy season.

Conc. Of Monocro-tophos (ppm)	Log of Conc.	Fish exposed	Fish dead	% Kill	Probit kill
20	1.30	10	01	10	3.71
25	1.40	10	02	20	4.15
30	1.48	10	05	50	5
35	1.54	10	07	70	5.52
40	1.60	10	08	80	5.84
45	1.65	10	08	80	5.84
50	1.70	10	10	100	8.09

Graph 15.1 Mortality of fish, *Nemacheilus botia* at different concentrations of monocrotophos for 14 days exposure period during winter season

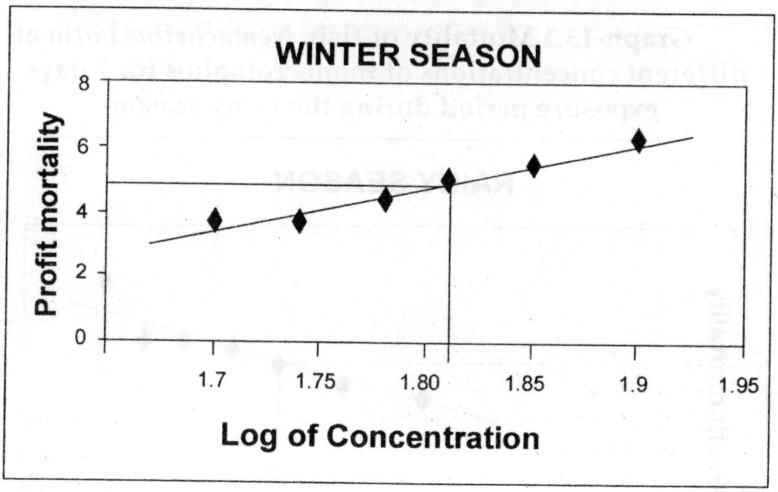

Graph 15.2 : Mortality of fish, *Nemacheilus botia* at different concentrations of monocrotophos for 14 days exposure period during summer season

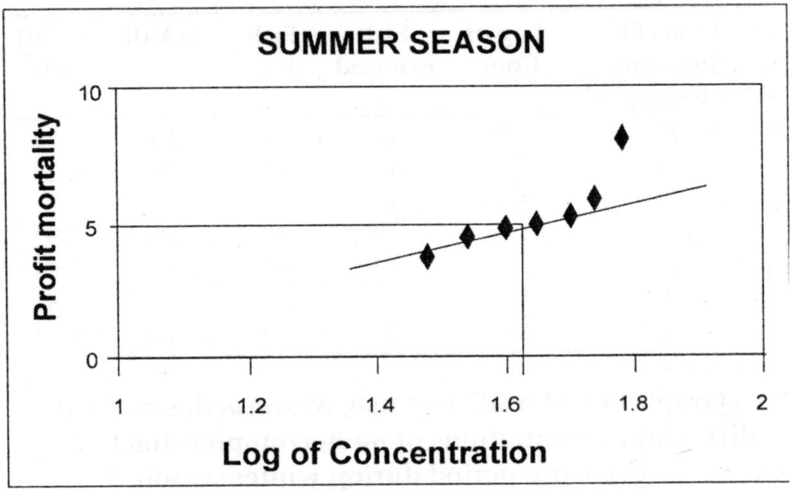

Graph-15.3 Mortality of fish, *Nemacheilus botia* at different concentrations of monocrotophos for 7 days exposure period during the rainy season

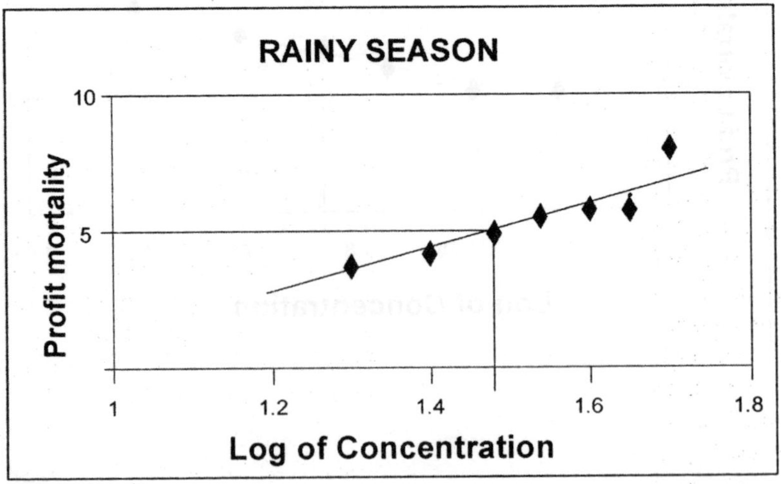

17

Effect of Heavy Metals on glycogen content of fresh water Prawn *Macrobrachium lamerrii*

B.S. Yadav
K.T. Patil

ABSTRACT

Heavy metals murcuric chloride and mercuric sulphate as a pollutant were used to see its effect on glycogen content of hepatopancreas and muscle of fresh water prawn's *Macrobrachium lamerrii.* The mature prawn were exposed to 0.12, 0.086, 0.062 and 0.032 ppm of mercuric chloride and mercuric sulphate. The LC50 values of 24, 48, 72 and of 96 hours respectively. After exposure periods the impact of acute concentration of heavy metals on glycogen contents were carried out. The results of the present investigation reveals a significant decrease in glycogen content of hepato pancreas however a variation in glycogen content of muscles were noticed after mercuric chloride and mercuric sulphate treatment.

Introduction

Prawns are recognized as valuable food source and form important diet of man has great nutritious value. In India mechanized fishing on modern line has been introduced in recent times and prawns are exported in thousands of tone fetching crores of rupees worth of foreign exchange. Rivers are exposed constantly to the newly developed industrial waste which includes heavy metals and heavy metals problem has been increasing day-to-day mainly due to industrial evolution. Most of the work on heavy metal pollution have been done on fishes. However there are few reports on heavy metals pollution in crustacean's Krishnagi, et. al. (1987). Therefore present investigation have been undertaken to the impact of heavy metals mercuric chloride and mercuric sulphate on glycogen content of fresh water prawn *M. lamerrii.*

Material and Methods

The fresh water prawn. *M. lamerrii* were collected from Girna Dam, Malegaon Dist. Nasik. India. Prawn's were reared in plastic troughs containing aerated tap water. The prawn's were acclimatized to the laboratory condition for 3-5 days prior the commencement of the experiment. Prawns were fed with pieces of earthworm and water was changed alternate and days for experiments approximately same size of animals were used water parameters such as Hardness (210 mg /l), pH. 6.8. and temp. (22°C) were recorded.

The stock solution of mercuric chloride (Manufactured by BDH laboratory chemicals, Glaxo laboratory, Bombay) and Mercuric sulphate (Manufactured by S.D. Fine Chemical Ltd. Boisar) was prepared by dissolving 1.0 gm in 250 ml of distilled water and from the stock solution different concentrations were prepared.

Prawns were exposed to lethal concentration of mercuric chloride and mercuric sulphate for 24 hours, 48, 72 and 96 hours. After exposure periods prawns were dissected and hepatopancreas and muscles were taken out for glycogen estimation (Kemp et. al. 1958).

Results

The results of the present experiments reveals that in

hepatopancreas a gradual decrease in glycogen content was obtained after acute exposure to mercuric chloride and mercuric sulpate. However in muscles an initial increase and then gradual decrease in glycogen content were obtained after mercuric chloride and mercuric sulphate exposure.

Table 16.1 : [a] Effect of mercuric cloride on glycogen contents of hepatopancreas and muscle of prawn *M. lamerrii*

Exposure period	Tissue Hepatopancreas	Muscle
Control	2.84 ± 0.42	1.38 ± 0.28
24 hrs	2.72 ± 0.38	1.82 ± 0.36
48 hrs	2.36 ± 0.22	0.98 ± 0.14
72 hrs	1.98 ± 0.22	0.66 ± 0.08
96 hrs	1.34.± 0.22	0.48 + 0.12

Table 16.1 : [b] Effect of mercuric sulphate on glycogen content of hepatopancreas and muscle of fresh water prawn *M. lamerrii*

Exposure period	Tissue Hepatopancreas	Muscle
Control	2.82 ± 0.38	1.38 ± 0.28
24 hrs	2.68 ± 0.42	1.74 ± 0.36
48 hrs	2.22 ± 0.32	1.26 ± 0.32
72 hrs	1.82 ± 0.18	0.84 ± 0.078
96 hrs	I .46 ± 0.22	0.4 ± 0.06

Discussion

Pollutant causes stress on the organisms and disrupt its metabolic activities. The biochemical changes in the organ's of animals exposed to pollutant have no definite pattern and physiological state of metabolic activity of an organism reflects in the utilization of their biochemical energy to counteract the toxic stress (Nagabhushanum et. al. 1987)

In the present experiment a significant ($P<.05$) decrease in glycogen content was observed in hepatopancreas of fresh water prawns *M. lamerrii* after acute exposure of mercuric chloride and mercuric sulphate. The decrease in glycogen content indicates its utilization for countering the heavy metals tress and its suggests prevalence of ancerobic conditions such as anoxia (Mayes 1977). Anoxia stress causes an increase in carbohydrate consumption this indicates that heavy metals mercuric chloride and mercuric sulphate may cause stress conditions similar to anoxia and it also leads to increased glycogenolysis. Decreases in glycogen in the tissues may be due to high energy demand during heavy metals exposure similar observation has been made by Jaiswal et. al. (1989).

Nagabhushanam, et.al. (1987) noticed a gradual decreases in glycogen content of hepatopancreas in fresh water prawn *M. kistnensis* after exposure to pesticide thiodon and phenthoate Machale, et. al. (1989) noticed similar results on crab *B. guerinii* after exposure to copper and cadmium. Khillare et. al. (1987) reported a significant decline in fresh water fish *channa gachua* on exposure to Lindane and Dimethoate. Further they stated that decrease in glycogen level may be due to toxic action on the tissues and the stored glycogen from liver might have utilized to counteract the pollutant stress Patel, et. al. (1991) reported a significant decrease in glycogen content during acute exposure to endosulfan on fresh water crab *Barytelphusa cunicularis*.

Reddy, et. al. (1986) observed a gradual decreases in glycogen contents in hepatopancreas of crab Oziotelphusa *senex senex*, further they reported that decline in glycogen content mignt be due to the enhancement of glucogenolysis by increase in phosphorelase activity.

In the present experiment when fresh water prawns were exposed to mercuric chlorides and mercuric sulphate, an initial increase and gradual decrease in glycogen content of muscles were observed. An increase in glycogen content during the first few hours may be due enchanced glyconeogenesis. Similar observation were made by Khan et. al. (1990) in fresh water crab *Barytelphusa cunicularis* after exposure to zinc sulphate and copper sulphate. Machale, et. al. (1991) obtained significant increase in glycogen content of crab *B. guerinii* after acute exposure to cuprous oxide. Similar observation were made by Shaikh (1996) on crab *B. guerinii* after exposure to mercuric chloride and meruric sulphate.

Above finding shows that pesticides alter the biochemical sequence of metabolism and the animal is able to adapt to the stress condition.

References

Jaiswal, K.B., R. Nagabhushanam and R. Sarojini (1989) Influence of Naphthalene on the tissues of prawn Macrobrachium kistnesis. *Geobios* 16, 197-202.

Kemp, A., and A.J.M. Kits, Van Heiginingen (1954) A colorimetric micromethod for the determination of glycogen of fishes. *Biochem.* J. 56.

Khan, A.K., R. Sarojini, P.R. Machale and R. Nagabhsuanam. (1990) Biochemical changes produced as a result of zine sulphate and copper sulphate in the muscle of fresh water crab. *Barytelphusa guerini.*

Killare Y.K. and S.B. Wagh (1987) Effect of Dimethoate and Lindane on tissue glycogen content in the fresh water fish *channa gachua*. *Proc. Nat symp ecotoxic* 91-93.

Krishanji A. P. M. S. Rege, and A. Joshi (1987) Toxic effect of certain heavy metals (Mercury, cadmium, Lacad, Arsenic and selenium) on the inter tidal crab Scylla serrata. *Mar. Environ. Res.* 21(2) 109-20.

Machale P.R., A.K. Khan, R. Sarojini and R. Nagabhushanum (1989) Copper and cadmium induced changes in Blood sugar level of the crab *Barytelphusa cunicalaris*. *Uttar Pradesh J. Zoot.* 9(1)113-15.

Machale, P.R., A.K. Khan and R. Sarojini (1991) Biochemical changes induced by cuprous oxide in the various tissues of the fersh water crab B. guerrini. *J. Adv zool.* 12(2)115-20.

Mayes, P.A. (197) Review of physiological chemistry. 16th. ed. Harper, H. A Rodwell, U. V. and Mayes P. A. Lange medical publications california. Nagabhushanum, R. Jaya Deshpande and R. Sarojini (1987) Effect of some pesticides on the biochemical constituents of fresh water prawn *Macrobrachium kistenensis*. *Proc Nat. symp. Ecotoxic.* 74.84.

Patel, R. T. (1991) Effect of endosulfan pesticide on some physiological aspecta of *Barytelphusa guerinii.* Ph.D. Thesis Marathwada University Aurangabad (17.5)

Reddy S.L. M. Venugopal, M. B. , R. K. and J. V. Ramana Rao (1986) In vivo effect metabolism in the tissue of a crab *Oziotelphusa senex senex*. Bull Environ. contain. Toxicol. 42-847.853.

Shaikh, I. S. (1996) Toxic effects of heavy metals on some physiological aspects of crab *Barytelphusa guerinii.* Ph.D. Thesis Marathwada University Aurangabad.

18

Impact of Cypermethrine pesticide to fresh water Crab, *Barytelphusa cunicularis*

B.S. Yadav
Shaikh I.S.

ABSTRACT

Acute toxicity of synthetic pyrethroids pesticide fenevalerate have been studied on fresh water crab *Barytelphusa cunicularis.* The LC50 values were found to be 0.2511, 0.2290, 0.1659 and 0.1548 ppm for 24, 48, 72 and 96 hrs. respectively. The highest LC50 values were recorded after 24 hrs. exposure and lowest at 96 hrs. exposure.

Introduction

Environmental pollution resulting from various factors particularly by the use of poisonous chemicals including pesticide, is matter of grave concern for us. In our country, the use of pesticide has been about 100,000 tonnes a year and areas of application cover both agriculture and public health. Though many workers have focused the light towards the toxic effect of pesticide on freshwater and marine crustaceans, such as Manikumar (1986), Avelin Mary, 1984, Jaya Deshpande 1986, Rao 1984, Martin, 1990, Yadav, 1991. But meagre literature is available on the toxicity of synthetic pyrethroid fenvalerate on the fresh water crustaceans. Therefore the present work has been undertaken to observe the toxic effect of fenvalerate on the fresh water crab *Barytelphusa cunicularis*.

Materials and Methods

The fresh water crab *Barytelphusa cunicularis* were procured from Girna Dam near Malegaon, Nasik (India). The crabs were reared in plastic trough containing tap water and acclimated to laboratory conditions for 3-4 days prior to the commencement of the experiment. Crabs were fed with the pieces of earthworm and water was changed after every alternate day. The physico-chemical characteristic of water such as pH-6.4, hardness 440 mg/litre and temperature 24°C were recorded.

The stock solution of pesticide fenvalerate was prepared by dissolving a known solution of pesticide in tap water and required concentration were prepared from the stock.

In this biostatic assay, mature crab of approximately same size were used. Each experiment repeated thrice and data was pooled for purpose of calculating the LC 50 values using finney probit analysis chart.

Results

The percentage mortality of crabs increased progressively up to 96 hrs. In all the concentrations of the pesticide fenvalerate, the LC50 values decreased with increasing the exposure period. In the mature crabs the LC50 values of fenvalerate were found to be 0.2511, 0.2290, 0.1659 and 0.1548 ppm after 24, 48, 72 and 96 hrs. exposure respectively.

From the above results it was observed that the LC 50 value found after 24 hrs. exposure was highest and LC 50 values of 96 hrs. was found lowest among all the exposure periods (Table 17.1).

Discussion

The physio-chemical and biological components of the environment play an important role in the manifestation of biological response to the pesticide (More 1969). Mortality in the crab is a more sensitive measure of toxicants. The present study of toxicity of fenevalerate to the fresh water crab, *Barytelphusa cunicularis* indicated that per cent survival rate of the rabs decreases with increasing, the concentration of pesticide.

In the present probe, acute toxicity test shows a relationship between the length of exposure period and concentration of pesticide. The LC50 values of the crab decreases gradually as the exposure period goes on increasing order. Acute toxicity involves the damage to the organism by fastest acting mechanism. Similar results were observed by the several workers such as Mirajkar 1984, Deshpande 1985, Manikumar 1985, Sharma 1989, Yadav 1992.

The results of the present experiment may be useful in evaluating the toxicity of fenevalerate pesticide to the marine and fresh water crabs which are an important food chain of the ecosystem.

Acknowledgement

Authors wish to thank to Hon. Prashant Dada Hiray, The General Secretary M.G. Vidyamandir, Malegaon for providing all laboratory facilities and inspiration during the course of this tasks.

Reference

Deshpande, J.R. (1985) Effect of some pollutants on the reproduction in crustacea. Ph. D. Thesis, Marathwada University, Aurangabad. M.S. India.

Finney, D.J. (1971) "Probit Analysis" Cambridge University Press, London.

Manikumar, D. (1986) Effect of pollutants on the marine prawn,

Penaeus merquinensis. Ph.D. Thesis Marathwada University, Aurangabad M.S. India.

Marry Avelin (Sr.) (1984) Effect of pesticide on physiology of the fresh water prawn, *Macrobrachium lamerrii*. Ph.D. Thesis, Marathwada University, Aurangabad (M.S.)

Martin, R.E. (1990) Impact of parathion on the Physiological response of fresh water prawn, *Caridina weberi*. Ph.D. Thesis, Marathwada University, Aurangabad, M.S. India.

Mirajkar, S.S. (1984) : Effect of pollutants on the neurosecretory system of the prawn. *Macrobrachium kistnensis*. Ph.D. Thesis, Marathwada University, Aurangabad, M.S. India.

Moore, N.W. (1969), Rural Medicine, 1:1.

Rao, K.S. (1984) : Effect of pesticide pollutants on reproduction and neurosecretion in marine edible crab *scylla Serrata*. Ph.D. Thesis. Marathwada University, Aurangabad, M.S. India.

Sharma, H.A., R. Sarojini and R. Nagabhushanam, (1989) : Physiological investigations of the Toxicity of pesticide methyl parathion and phosalone to the commercial crab *Scylla serrata*. *Environ. State and Bio-Science* 1967-98.

Yadav, B.S., R. Sarojini, R. Nagabhushanam, Toxic effect of endosulfan on fresh water prawn *Caridina weberi*. *J. Advance. Zool* 12(1) 19-22.

Table 17.1 : Relative toxicity of fenvalerate when crabs were exposed to the acute concentration (24 hrs. to 96 hrs.)

Exposure Periods	Regression equation	LC50 + SE	Homogenicity Heterogenicity	Fiducial Limit M1	Fiducial Limit M2	Relative toxicity
24	y=3.14 + 4.5 x	.2511+.584	- 1.95	0.2465	0.4757	1
48	y=3.69 + 3.9 x	.2290+.1250	4.9176	0.2147	0.2433	1.0965
72	y=4.17 + 3.5 x	.1659+.1779	0.1943	0.03125	0.3005	1.5135
96	y=4.47 + 4.05 x	.1548+.054	-0.8162	0.0372	0.2724	1.6220

19

Histopathological studies on intestine of *Columba livia* Gmelin, 1789 Infected with Cestode Parasites

S.D. Patil
Hemlata S. Chaudhari (Wankhede)

ABSTRACT

The cestode parasites *Cotugnia aurangabadaensis* Shinde, 1969 and *Paruterina sp.* found infected to the avian host *Columba livia* Gmelin, 1789. Histopathological studies have been made to assess the extent of damage caused by the parasites. It includes destruction and extrusion of intestinal villi, inflammatory fibrosis due to cysts. The scolex of *Cotugnia aurangabadensis* in non-penetrative type while *Paruterina sp.* is penetrative type. The extent of damage is proportional to the penetration of scolex. Cysts found encircled with connective tissue sheath deep in the submucosa.

Keywords: *Cotugnia aurangabadensis*, Histopathology, *Paruterian sp. Columba livia* Crypts of Lieberkhun.

Introduction

In Cyclophyllidea, many genera like *Dipyllidium canium, Cotugnia bhaleraoi, Raillietina R. tetragona, Echinococcus granulossus* are studied for histopathology, host-parasite relationships. Some scolec were penetrative type and others were non-penetrative type (Shinde and Mitra 1980). In penetrative type, attachment is very intimate and Crypts of Lieberkhune are invaded while in non-penetrative type it superficially attached to mucosal epithelium of intestinal villi. Important contributions in this direction were made by Tuli, 1992; Mitra and Shinde, 1981; Jadhav and Shinde, 1981; Lakshma, Reddy, et. al., 2006; Banarjee, et. al., 2007.

In the present study an attempt has been made to visualize the histopathological changes that are caused to the intestine of avian host *Columba livia,* Gmellin, 1789 due to infestation of cestode parasite *Cotugnia aurangabadensis* Shinde, 1969 and *Paruterina sp.*

Material and Methods

Intestines of host bird *Columba livia* were examined and observed to see the degree of infection. The worms which attached to the intestine were kept intact and small pieces of intestine were fixed in Bouin's fixative, washed thoroughly. Dehydrated through ascending alcohol grades, cleared in xylene and embedded in paraffin wax (M.P. 52°-54°C). The transverse and longitudinal sections were cut on rotary microtome and eosin used as counter stain.

Description

The changes that occur in the intestine of host are due to mechanical damage and may be due to the release of toxins by the parasites. Intestine is one of the important parts of the alimentary canal concerned with several functions like chemical digestion and absorption of nutrients. The histopathology caused extensive damage to various layers of intestine right from epithelium of mucosa to the muscularis mucosae.

The infected intestine of *Columba livia* by *Cotugnia aurangabadensis* shows scolex attached to the intestinal villi through large rostellum. It is non-penetrative type of scolex as worm invades only the villi but not the crypts of Lieberkhun. Mature and gravid proglottids found freely sus-

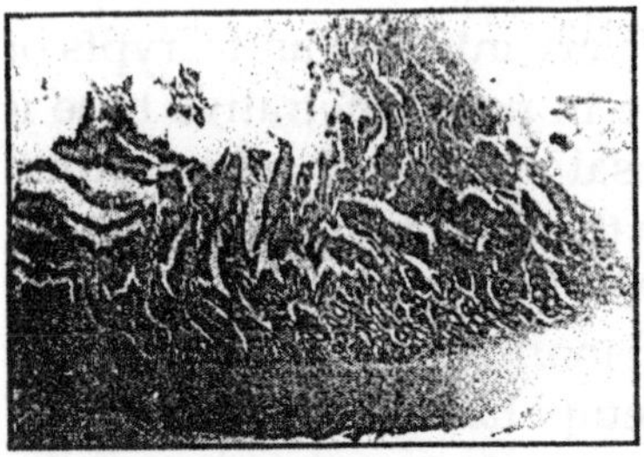

Fig. A : Non-infected intestine of ***Columba livia***

Fig. B : Scolex ***Cotugnia aurangabadnesis*** attached to intestinal villus

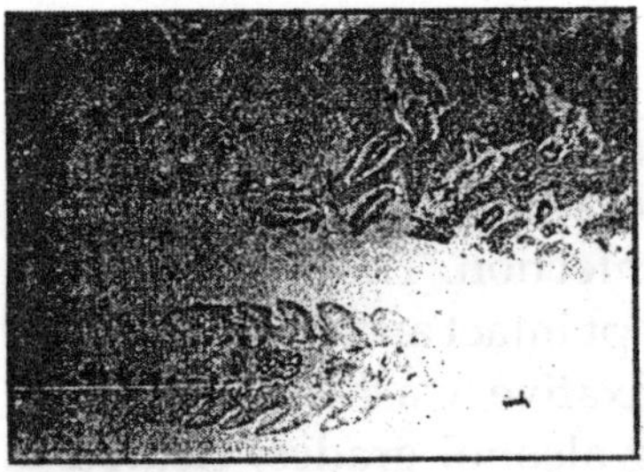

Fig. C : Mature proglottids of ***Cotugnia aurangabadnesis*** in lumen of intestine

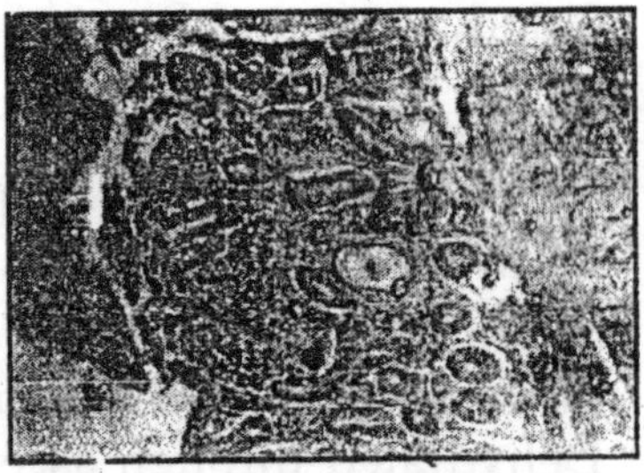

Fig. D : Cyst of *C. aurangabadensis* in subsmucosa layer

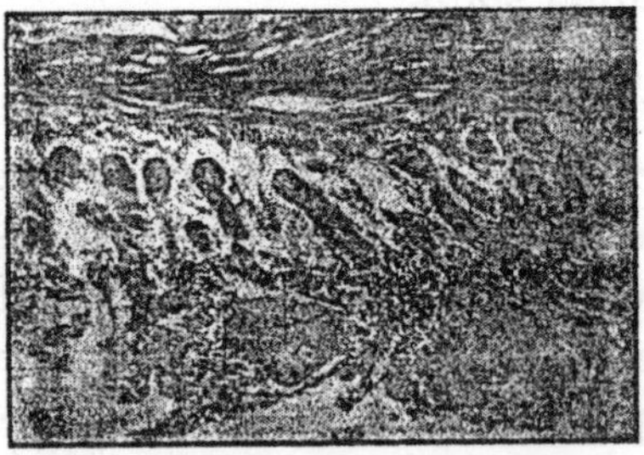

Fig. E : Scolex of ***Paruterina shindel*** n.s.p. approching to the Crypts of Lieberkuhn.

Fig. F: Cyst of ***P. shindei*** in Submucosa and Mature Proglottids among intestinal villi of ***Columba livia***

pended in the lumen of intestine. Cysts found deep in the submucosa, just above the muscularis externa and encircled with connective tissue sheath.

The infected intestine of blue rock pigeon by *Paruterina sp.* is heavily destroyed, rupturing the villi and reaching deep up to submucosa. It is penetrative type of scolex. Mature proglottids found among the intestinal villi. A pair of cyst shows fibrosis deep in the submucosa. It leads to lesion on the intestine causing pores and bleeding. Gravid segments freely suspended in the lumen of intestine.

Discussion

Various helminth parasites shows pathological consequences of parasitic effects on birds are well documented. Cestode parasites influences the avian health, causes morbidity and also mortality which pose a major threat to avian population. However, the extend of damage depends upon depth of penetration of scolex, type and number of cestode parasite and site where they localize in the body of host (Paperena and Zwerner, 1979).

The infected intestines were looked diseased with swellings blood clot, all along the alimentary canal and bleeding at certain places. The intestine infected with *C. Aurangabadensis* is superficial and not invades the Crypts of Lieberkhun, so the worm is non-penetrative type (Chincholikar, 1978, Joshi , 1971). Tissue reaction and cellular infiltration occurs. It is negligible when parasite attached superficially with non-penetrative type of scolex (Mitra and Shinde 1990). But become extreme with extensive granuloma and fibrosis when the scolex is attached to the submucosa or entirely perforates the wall of intestine (McDonough and Gleason 1981, Kobate 1985).

Histopathology revealed disseminated erosion at the site of attachment, lymphocyte migration and hyperplasia of connective tissue in the submucosa (Ivona M. Ladenea 2006).

Conclusion

Cotugnia aurangabadensis has non-penetrative scolex as attaches superficially to the intestinal villi but causes fibrosis, tissue reaction and cellular infiltration. This inter-relation of the parasite with the host results in the survival of the parasite and a slight damage to the host intestine as the parasite is non-penetrative type.

The *Paruterina sp.* are penetrative type and it pierces through crypts of Lieberkhun, mucosa and up to the submucosa layer. It pose a serious threat to the birds. However, the extend of damage or pathogenic conditions depends on number of invasive parasites and the site where they localize in the host body.

Acknowledgement

Author thanks Head, Department of Zoology, Dr. Babasaheb Ambedkar Marathwada University, Aurangabad (Mah) for providing necessary laboratory facilities.

References

Reddy, P. Lakshma, G. Banarjee, G. Rajender, V. Bikshapthi, M. Swami (2006). Histopathological and Histochemical abnormalities induced *Euclinostomum heterostomum* in the liver of fresh water fish *Channa punctatus*. *J. Aqua. Biol.* vol. 21 (2) 2006 : 263-267.

Chincholkar, L.N. and G.B. Shinde. (1956) Studies on morphology of *Cotugnia celebensis* Yamaguti (Cestoda : Davaineidea) Part IV : Histology. *Acta. Para. Pol.* Poland. 1956.

G. Banarjee; B. Laxma Reddy and Bikshapti (2006) : Histopathology and Histochemistry of the intestine of *Clarias batrachus* due to trematode *Genarchopis goppo* (Ozaki, 1925). *J. Aqua. Biol.* vol. 21(2):257-62

G.B. Shinde and K.B. Mitra (1980). On the host parasite interaction of *Raillietina (Raillietina) tetragona* in *Gallus domesticus* at Aurangabad, India. *Biology* Vol. II, no. 2 : 31-34.

Ivona M. Ladena (2006). *Hepatoxylon trichuri* (Cestoda; Trypanorhyncha) Plerocercoids in cage-reared northern blue fin tuna *Thunnus thunnus* (Ostiecthyes : Scoimbridge). *Acta. A driat.* 47(1) : 79-83.

Joshi, S.C., S.K. Kamalpur (1971). Occurrence of *Cotugnia bhaleraoi* (Madaliar, 1943) in desifowl in M.P. and some observations on its histopathology. *Int. Vet. J.*

Mitra, K.B. and G.B. Shinde (1981). Attachment of *Davainea sp.* to the villi of *Gallus domesticus.* Bioresearch 1981, 5(1) : 89-90. Luka S.A. and I.S. Ndam (2007). Gastrointestinal parasites of chickens. *Gallus gallus domesticus* (Linnaeus, 1758) in Samaru, Zaria, Nigeria. *E.World J.* Vol. 2(1), SWJ : 27-29.

McDonoug, J.M. and L.M. Gleason (1981). Histopathology in rainbow derter *Etheostoma caeouleum* resulting from infections with the acantocephalons *Pomporhynchus bulbocolli* and *A. dinu. J. Parasitol* 67 : 403-07.

Paperna, I. and D.F. Zwerner. (1976). Parasites and diseases of stripped Bass *Morone sexatalic* (Walbaum) from the lower Chesepeake bay. *J.Fish Biol.* 9 : 267-287.

Tuli, J.S.; H.S. Bali., P.P. Gupta (1976). Histopathology and Histochemical Studies of intestine in Cestodiasis of poultry. *Ind. J. Comp. Microbiol. Immun. and Infc. Diseases* 13 (3 and 4) : 100- 04.

20

Effect of Nuvan Pesticide on Haemocytes of Beetle *Mylabris pustulata*

K.T. Patil
B.S. Yadav

ABSTRACT

In the present study, Nuvan pesticide was used as pollutants to study its impact on morphology and number of Haemocyte in haemolymph of beetle *Mylabris pustulata.*

The beetles were exposed to 6ppm, 4, 3 and 2ppm of Nuvan, the Lc50 value of 24 hrs, 48, 72 and 96 hrs respectively. After exposure periods the effect of Nuvan on haemocytes were carried out. The result of the present investigation reveals that the haemocytes changed its form and appear irregular in shape due to effect of Nuvan pesticide which causes adverse effect on haemocytes number. The total number of haemocytes is found to decreases.

Introduction

Insects possess an open circulatory system with the haemolymph occupying the general body cavity, the haemocoel. Haamocoel in insect is often divided into three major sinuses, a dorsal pericardial sinus, a middle perivisceral sinus and a ventral perineural sinus. The haemolymph circulates round the body, bathing the tissues directly. It consists of fluid plasma present in haemocoel and floating in plasma is haemocyte. Insect haemolymph is not entirely analogus to vertebrate blood.

Jones (1964) has classified the haemocytes in four types viz., prohaemocyte, plasmocyte, granular haemocyte, cystocytes. In the present studies, efforts have been made to study the morphology and number of haemocyte in haemolymph of "Blister beetle". *Mylabris pustulata* in relation to pesticide exposure.

Nowadays our environment is highly contaminated due to indiscriminate use of pesticide in agriculture sector, which has leads to deleterious effect on non-target species. Hence an attempt has been made to study the effect of organophosphate pesticide the Nuvan on haemocytes mainly on the structure and number. The insecticide Nuvan was sprayed on the flowers and then observed the mortality and survival of the beetle after 24 hrs. Followed by observation on, the effect of Nuvan on the haemocytes. The total number of haemocytes in an insect showed that small insect have fewer haemocytes than larger insects. Number of haemocytes varies from species to species of insects.

MATERIAL AND METHODS

Site of Collection

Blister beetles were collected from the Dabhadi area near Malegaon, Nasik district. Blister beetles were collected in cage along with Hibiscus flower and brought to the laboratory.

Rearing of Blister Beetle

The blister beetles were maintained in cage containing flowers. The flowers were changed everyday. The live, healthy adult Beetles were selected for haemocyte study.

Haemocyte Mounting and Staining of Haemocytes

The haemolymph was taken by direct microinjection syringe and placed on clean slides, spread by narrow edge of another slide like blood smear preparation. The haemolymph film was air dried for 20-30 minutes and fixed in methanol: acetic acid (3:1) fixative for 15 minutes. Then the slides were dried and stained with Giemsa (2%) for 3-5 minutes. After staining, slides were washed in distilled water and air dried. The air dried slides were dehydrated, cleared in Xylene and mounted in DPX mountant. The slides prepared in this way were scanned first under lower power and then under high power of microscope for morphological changes. Selected haemocytes were microphotographed under high power.

Treatment of Insect

- The healthy Blister beetles of approximately same size were selected for the experiment.
- The 2 ppm to 6 ppm Nuvan solutions were sprayed on the flowers and such a sprayed flowers were kept in cylinder containing beetles.
- Observations were made for mortality and survival of the beetles after 25 hours.
- LC50 values were calculated using Finney's method (1971) following empirical probit method.

Haemocyte Count

- Haemolymph was sucked into W.B.C. pipette upto 0.5 mark and immediately diluted with diluting fluid by sucking up to the mark.
- The pipette was thoroughly shaken for complete mixing of haemolymph and diluting medium.
- It was allowed to stand for 2 to 3 minute, first two drops were discarded and then the two drops of this diluted haemolymph was placed on counting chamber and

then covered with clean cover slip without air bubble.

- Slide was observed under low magnification of microscope and then under high magnification.
- Haemocytes were counted, number was recorded and total haemocyte count was calculated.
- For mounting of haemocytes of Nuvan treated beetles, the method adopted as normal haemocyte mounting was carried out. The slides prepared was scanned under microscope and micro-photographed.

RESULT AND OBSERVATION

Morphology of Normal Haemocytes

1. **Prohaemocyte :** These cells are oval in shape; large with centrally located nucleus the cytoplasm occupies a narrow area around the nucleus.

2. **Palsmatocytes :** They are variable in shape and contain moderate amount of rough endoplasmic reticulum and Golgi complex.

3. **Granulocyte :** They contain large amount of endoplasmic reticulum and Golgi complex. These cells contain abundant granules in cytoplasm.

4. **Spherule Cells :** These cells are characterized by large, retractile spherules that occupy 90% of the cytoplasm.

5. **Oenocytoid :** Oenocytoid cells are largest among the haemocytes and show little development of rough endoplasmic reticulum and Golgi complex.

6. **Adipohaemocytes :** These cells are irregular in shape, contain variable number of lipid droplets.

Observation of treated haemocytes showed morphological changes like

(a) Haemocytes, turn into drum shaped bodies.

(b) Nuvan intoxication caused disintegration of haemocytes.

(c) All types of haemocytes showed dark nucleus.

A large number of dot shaped particles were observed in haemocyte cytoplasm.

(d) Haemocytes become irregular in shape due to cytoplasmic extensions.

HAEMOCYTE COUNT

Normal Haemocytes/mm^3 = 28650 haemocyte/mm^3
Treated haemocyte/mm^3 = 23900 heamocyte/mm^3

Discussion

The volume of haemolymph often falls down to its minimum level after mounting. It fluctuates daily in relation to feeding, flight and physiological conditions. The pH of haemolymph is usually between 6.4 to 6.8. The reported counts of the number of haemocyte per cubic millimeter of haemolymph show considerable variation (Lange, 1932). Haemocytes in cricket *Gryllus assimilis* ranged from 15,000/mm^3 to 2,75,000/mm^3.

Differential counts of haemocytes of different insects reported in the literature show that comparison among them is impossible due to differences in techniques used. But, the total and differential counts appear to influenced some what by the stage of development (Tauber & Yeager, 1934, 1935, 1936).

Rowely and Ratcliffe (1981) and Gupta (1979a, 1979b, 1985) explain six main types of haemocytes viz-prehaemocyte, plasmocytes, granulocyte, spherule and oenocytoid, adipohaemocytes. Number of haemocytes in haemolymph of insects varies from species to species. In a adult female Anopheles mosquitoes—10,000/mm^3, adult periplaneta—90,00,000/mm^3, *Gryllus assimilis*—15,000 to 2,75,000/mm^3. The volume of blood may vary at different stage in the life cycle and according to the physiological conditions of insect (Lee, 1961; chapman, 1972).

Intoxication or infection causes marked changes in haemocytes i.e. rapid decline in the haemocyte count which returns to its original level within an hour. The sudden reduction in number of haemocytes is largely as result of removing the plasmocytes from circulation. The responses of haemocytes in pupal lepidoptera is quite different during infections stages (Chamman, 1972)

References

Chamman, R.F. (1972) The Insects Structure and function. ELBS, London 819, pp. 106-121.

Finney, D.J. (1971) probit analysis 3rd Ed. Cambridge University, London.

Gupta, A.P. (1979) Haemocyte type : Structure Synonymies Inter relationships and toxemic significance In. Insect haemocytes. (OD, A.P. Gupta), pp. 85-128.

Gupta, A.P. 1979 Indentification key for homocyto type in hanging drop preparation. In: Insect haemocytes (OD. A.P. Gupta) pp. 5 : 27-52 Cambrian Uni. Press Cambridge.

Jones, J.C. 1964 the circulatory system of insects. P. 1-107, In M.R. Rockstein. (ed) Physiology of Insect. Vol. 3. Academic Press, London.

Lee, R.M. (1961), The Variation of blood volume with age in the desert locust Schitocerca gregari. J. Insect. Physioloy 6:35-51

Apendix I

Indian Standard for Drinking Water

The safe drinking water could be defined as the water which is free from pathogenic micro-organisms, poisonous substances, excessive amounts of minerals and organic matters which would produce undesirable physiological effects. It should be free from odour, turbidity, taste and odour on moderate temperature and aerated as well as non-corrosive and non-scale forming.

The establishment of minimum standards of quality for public water supply is of fundamental importance in achieving this ideal. Standards of quality from the yardstick with which the quality control of any public water supply has to be assessed. In India, certain minimum standards have been prescribed and given here.

Physico-Chemical Characteristics

Substances and characteristics affecting the acceptability of water for domestic use the physico-chemical quality of drinking water should not exceed the limits given below :

Substances or characteristics	Undesirable effect that may be produced	Highest Desirable level	Maximum Permissible level
Substances affecting the colour, TCU(I)	Discolouration	5 units	25 units
Substances causing odour		Unobjectionable	
Taste		Agreeable	
Turbidity NTU(2)	Gastrointestinal irritation	5 units	10 units
Dissolved solids	Gastrointestinal irritation	500 mg/L	2000 mg/L
pH range	Taste, Corrasion and scale formation	6.5 to 8.5	No relaxaiton

Alkalinity	Unpleasant taste	200 mg/L	600 mg/L
Total hardness	Taste and scale formation	300 mg $CaCO_3$/L	600 mg $CaCO_3$/L
Calcium	Taste and scale formation	75 mg Ca/L	200 mg Ca/L
Magnesium	Taste, scale, formation and gastrointestinal irritation in presence of sulphate	30 mg Mg/L	100 mg Mg/L
Copper	Astringent Taste, discolouration corrosion of pipes fitting and utensils	0.05 mg Cu/L	1.5 mg Cu/L
Aluminium	Cumulative effect is reported to cause dementia	0.03 mg/Al/L	0.2 mg/Al/L
Boron		1 mg B/L	5 mg B/L
Iron	Astringent taste, discolouration, turbidity and deposits growth of iron bacteria in pipes	0.3 mg Fe/L	1.0 mg Fe/L
Chloride	Taste, palatability and corrosion in hot water systems are affected	250 mg Cl/L	100 mg Cl/L
Residual chlorine		0.2 mg Cl/L	
Sulphate	Gastrointestinal irritation when combined with manganese or sodium	200 mg SO_4/L	400 mg SO_4/L
Nitrate	Infantile methaemoglobinaemia	45 mg NO_3/L	100 mg NO_3/L
Fluoride	Fluorosis	1.0 mg F/L	1.5 mg F/L
Anioxaic detergents	Floth in water	0.2 mg MBAS/L	1.0 mg MBAS/L
Phenolic compounds	Objectionable taste and odour	0.001 mg C_6H_5OH/L	0.002 mg C_6H_5OH/L
Maganese	Astringent taste, discolouration, turbidity and deposits in pipes	0.1 mg Mn/L	0.3 mg Mn/L
Zinc	Astringent taste and opalescence in water	5.0 mg Zn/L	15 mg Zn/L
Chromium	Carcinogeric	0.05 mg Cn/L	No relaxation

Lead	Toxic	0.05 mg Pb/L	- do -
Cadmium	- do -	0.01 mg Cd/L	- do -
Arsenic	- do -	0.01 mg As/L	- do -
Mercury	- do -	0.001 mg Hg/L	- do -
Cyanide	- do -	0.005 mg CN/L	- do -
Selenium	- do -	0.01 mg Se/L	- do -
Polynuclear aromatic hydrocarbon	Carcinogenic	—	—
Mineral oils	Undesirable taste and odour after chlorination	0.01 mg/L	0.03 mg/L
Pesticides	Toxic	Absent	0.01 mg/L
Radioactive materials			
(I) α-emitters			0.1 Bq/L
(II) β-emitters			1.0 pci/L

Biological Characteristics

(i) Virological Examination

Virus of fecal origin such has enteroviruses, revoviruses, adneoviruses A., rotaviruses, echoviruses, polioviruses, coxsackieviruses and hepatitis viruses should not be present in drinking water. In practice, 0.5 mg/L of free chlorine for one hour is sufficient inactivate viruses in drinking water.

(ii) Bacteriological Examination

(a) Coliform count in any sample of 100 ml should be zero. A sample of water entering the distribution system that does not conform to this standard calls for an immediate investigation into both the efficacy of the purification and method of sampling.

(b) Water in the distribution system should satisfy all the three criteria indicated below :

- *E. coli* count in 100 ml of any sample should be zero.
- Coliform organisms not more than 10/100 ml should be present is any sample.
- Coliform organisms should not be detectable in 100 ml of any two consecutive samples.

(iii) Examination of Microscopic Organisms

The drinking water should be free microscopic organisms as algae, zooplanktons, flagellates, parasites and toxin-producing organisms. Amoebiasis, guinea-worm disease, schistosomiasis, algal toxicity and parasitic infections are commonly caused by biologically contaminated water.

References

BIS (2003) Indian Standard of Drinking water - Specification (First Revision).

Bureau of Indian Standards, New Delhi.

Appendix II

General Standards* for Discharge of Environmental Pollutants

Part A : Effluents

S.No.	Parameter	Inland Surface water	Public Sewers	Land for irrigation	Marine/coastal areas
1	2	3	4	5	6
1.	Colour and odour	See 6 of Annexure-II	-	See 6 of Annexure-II	See 6 of Annexure-II
2.	Suspended solids mg/l, max.	100	600	200	(a) For process waste water (b) For Cooling water effluent 10 per cent above total suspended matter of influent
3.	Particle size of suspended solids	shall pass 850 micron micron IS Sieve	-		(a) Floatable solids solids max. 3 mm (b) Settleable solids, max 856 microns
4.	pH value	5.5 to 9.0	5.5 to 9.0	5.5 to 9.0	5.5 to 9.0
5.	Temperature	shall not exceed 5°C above the receiving water temperature	-	-	shall not exceed 5°C above the receiving water temperature
6.	Oil and grease mg/l max.	10	20	10	20

7.	Total residual chlorine, mg/l max	1.0	--	--	1.0
8.	Ammonical nitrogen (as N); mg/l, max.	50	50	--	50
9.	Total Kjeldahl nitrogen (as N); mg/l, max. mg/I, max.	100	--	--	100
10.	Free ammonia (as NH3) mg/l, max.	5.0	--	--	5.0
11.	Biochemical oxygen demand (3 days at 27°C), mg/l, max.	30	350	100	100
12.	Chemical oxygen demand mg/l, max.	250	--	--	250
13.	Arsenic (as As)	0.2	--	0.2	0.2
14.	Mercury (As Hg) mg/l, max.	0.01	0.01	--	0.01
15.	Lead (as Pb) mg/l, max	0.1	1.0	--	2.0

16.	Cadmium (as Cd) mg/l, max.	2.0	1.0	--	2.0
17.	Hexavalent chromium (as Cr^{6}), mg/l max.	0.1	2.0	--	1.0
18.	Total Chromium (as Cr) mg/I, max.	2.0	2.0	--	2.0
19.	Copper (as Cu) mg/l, max.	3.0	3.0	--	3.0
20.	Zinc (as Zn) mg/l, max.	5.0	15	--	15
21.	Selenium (as Se)	0.05	0.05	--	0.05
22.	Nickel (as Ni) mg/l, max.	3.0	3.0	--	5.0
23.	Cyanids (as CN) mg/l, max.	0.2	2.0	0.2	0.2
24.	Fluoride (as F) mg/l, max.	2.0	15	--	15
25.	Dissolved Phosphates (as P), mg/I, max	5.0	--	--	--
26.	Sulphide (as S) mg/l, max.	2.0	--	--	5.0
27.	Phenolic compounds (as C_6H_50H) mg/I, max.	1.0	5.0	--	5.0

28.	Radioactive materials :				
	(a) Alpha emitters micro curie mg/l, max.	10^{-7}	10^{-7}	10^{-8}	10^{-7}
	(b) Beta emitters micro curie mg/l	10^{-6}	10^{-6}	10^{-7}	10^{-6}
29.	Bio-assay test	90% survival of fish after 96 hours in 100% effluent	90% survival of fish after 96 hrs. fish after 96 hrs.	90% survival of fish after 96 hrs. fish after 96 hrs.	90% survival of fish after 96 hrs. fish after 96 hrs.
30.	Manganese	2mg/l	2mg/l	--	2 mg/l
31.	Iron (as Fe)	3 mg/l	3 mg/l	--	3 mg/l
32.	Vanadium (as V)	0.2 mg/l	0.2 mg/l	--	0.2 mg/l
33.	Nitrate Nitrogen	10 mg/l	--	--	20 mg/l

* These standards shall be applicable for industries, operations or processes, other than those industries, operations or process for which standards have been specified in Schedule of the Environment Protection Rules, 1989.